P9-DJV-515

THE GEOMETRY OF VIOLENCE AND DEMOCRACY

THE GEOMETRY OF VIOLENCE AND DEMOCRACY

Harold E. Pepinsky

INDIANA UNIVERSITY PRESS
Bloomington and Indianapolis

The paper used in this publication meets the minimum requirements of American National Standard for Information Sciences—Permanence of Paper for Printed Library Materials, ANSI Z39,48-1984.

∞™

Manufactured in the United States of America

Library of Congress Cataloging-in-Publication Data

Pepinsky, Harold E.
The geometry of violence and democracy / Harold E. Pepinsky.
p. cm.
Includes bibliographical references.
ISBN 0-253-34343-7 (cloth)
1. Violence. 2. Crime. 3. Democracy. 4.Nonviolence.
I. Title.
HM281.P385 1991
303.6—dc20 90-4704
CIP

1 2 3 4 5 95 94 93 92 91

To Mama and Daddy,
Norwegian friends,
and the Church of the Earthborne

CONTENTS

FOREWORD

Violence is the antithesis of democracy. Such is the clear and present revelation of Hal Pepinsky's essays. And punishment—pursued in the name of virtue—is the further perpetuation of violence, as well as another step backward from the possibility of democracy. All the acts of crime follow from the domination of some over others—whether in the exercise of political power, economic exploitation, cultural hegemony, or spiritual authority. The solutions to crime within these systems of domination only serve to continue the problems they seek to eliminate

Democracy is alive and well as an ideal and as a reality to be achieved. The call was evident in Tiananmen Square. Each day we hear the struggles for democracy in countries around the world. And at the same time we learn of the violent efforts to still these moves. The dynamic of our human history is at this moment the struggle for democracy. This time the movement is not for the liberal democracy that isolates people from one another, but for a socialist democracy that seeks the equality and unity of all people, within their diversity; a democracy that is based on care and generosity and equal distribution of resources; a democracy of respect and loving-kindness. This is a democracy that follows the principle of the universe: nature as an original state of unity, an interconnection of all things and all beings.

As we are reminded in this wonderfully original and compassionate book before us, democracy begins in our daily lives, within ourselves and within our families, with our children, and spreads to the structures we create, and comes home to us again to inform and guide our lives. The implications of this for the practice of criminal justice in the United States are revolutionary. The author notes that his aim is modest, though: that a few readers may find it possible to take a few steps to think and act their way out of the crime and violence surrounding us. There is no place in our daily lives where democracy is not appropriate.

Let us read these words before us with an open mind and an open heart. Let us practice daily the wonder of our mutual existence. With this awareness, as in the Zen search for the ox, we go to town with helping hands. That we may live once again in the light of our true nature, democratically.

Richard Quinney

ACKNOWLEDGMENTS

I recognized only recently that my mother, Pauline N. Pepinsky, has been a pacifist throughout my life. All the signs were there, and her influence on me has been manifest. My father, Harold B. Pepinsky, and my mother raised me to think critically and cross-culturally about social problems. I am grateful to them for starting me on the intellectual quest that has led to this volume.

Chapter 2 begins with a return visit to Norway. The friends I have met there, who have taught me so much about living peacefully, are too numerous to name. Readers will see that I draw heavily on the work of Nils Christie and Birgit Brock-Utne. Two Norwegian friends, Lill Scherdin and Per Ole Johansen, have taken special pains to review and talk over my ideas with me. Many other Scandinavians have contributed to my thinking in important ways, too.

The Church of the Earthborne is a group that my wife, Jill Bystydzienski, my daughter, Katy, and I belong to in the Bloomington area. We have worked and struggled together for a few years now, trying to build a community of peace and harmony with the earth that supports us. Being out in the woods with these good people has more than once restored my sense of sanity in a violent, crazy world. One family in this group—Bill, Glenda, Denise, and Dietrich Breeden—is featured in chapter 4.

My parents, Norwegian friends, and the Church of the Earthborne represent special support and inspiration I have received from birth to the present. This book is dedicated to them.

They represent only a fraction of those who have sustained me and helped me to understand violence and democracy. My closest partners and teachers for more than a decade have been Jill and Katy, to whom I have dedicated earlier books. Chapter 7 particularly reflects the love and understanding they have given me. I dedicated another book to Les Wilkins, and this work reflects the importance he as a criminologist places on building "democracy." Richard Quinney has honored and flattered me by contributing a foreword, and by collaborating on editing a book of original readings concerning criminology as peacemaking, which IU Press is publishing as a companion volume to this one.

I want especially to acknowledge the importance of the colleagueship I enjoy in the Department of Criminal Justice where I teach. My colleagues, and indeed my students, have opened new vistas to me and have taken pains time and again to help me work through ideas of how violence and democracy operate. It is an unusually cosmopolitan department, where interdisciplinary and cross-cultural inquiry is prized and nurtured. I feel very lucky to be there.

For review and helpful suggestions for revision of an earlier draft of this work, I thank John Smylka. Gina Doglione and Jackie Moore have done a lot to help get this manuscript together. And without Judy Kelley's office support, the department wouldn't function. I thank them all for their generous and skilled assistance.

I hope the many others who have contributed so much in so many ways to my thinking, and helped me so much in my work, will not feel slighted that I fail to mention them all by name. It is an imposing task indeed to thank all those who have contributed to a work which is not a narrow, specialized inquiry but an integral part of the living and understanding of my daily existence in its entirety. Many of you are cited and named in the text that follows.

Although the responsibility for any problems of expression is mine, I am more than ever aware that a book is not the property or creature of its author. Whatever worthwhile comes out in this book is a gift of experience and understanding that at most passes through me as a gift from others. My heartfelt thanks and appreciation go to all of you who have made this book possible.

THE GEOMETRY OF VIOLENCE AND DEMOCRACY

1

Introduction

This book is part of an intellectual odyssey. For more than twenty years, I have sought numerous times in numerous ways to discover what "crime" is.

In my second year of law school, I took my first criminology course. The year was 1967. The reports of the President's Commission on Law Enforcement and Administration of Justice were hot off the press. My teachers were the executive director of that commission, James Vorenberg, and the director of the Task Force on Assessment of Crime, Lloyd Ohlin. Ohlin had commissioned and published the results of the first three large-scale victim surveys ever done. These studies concluded that crimes were vastly underreported to police. The *Task Force Report on Assessment of Crime* we were assigned to read gave the victim survey results, then proceeded to use police data to show that crime was rapidly increasing in the turbulent sixties.

I was astounded. First the report claims that police figures are wildly inaccurate, then it relies on police figures to describe the nation's crime problem. I said so in class. I asked Lloyd Ohlin, "How do we know that police figures have anything at all to do with what's happening on the streets?" Lloyd, who is a gentle soul and a very accomplished criminologist, replied, "We don't know, but they're the best we have to work with."

American law school courses generally have but one final exam. The classic question is to spot the legal issues in a hypothetical situation. Here the situation was to imagine oneself an aid to a member of Congress who intended to take a stand on massive appropriations to fight crime. I was to advise the member what the state of the crime problem was. I wrote that there was no reason to believe the police figures, and that I wasn't even so sure about relying on the victim survey figures—that for all we knew crime was no worse than it ever had been, that we had no basis for deploying more resources, and so we ought to save the taxpayers' money.

I got one of my two C's in law school for that exam. I think that did it. There's a rebel in me that regarded that grade as a challenge. I had to find out which one of us was right.

For my dissertation I rode 600 hours in police squad cars in the high-crime area of Minneapolis, carefully logging data on every call the police got, and on whether they reported an offense as a result. Here, surely, was the physical nexus between police figures and what happened on the street.

The trouble was that my data showed me that police reporting was almost entirely determined by simple rules of police organization which had nothing whatsoever to do with what was happening on the street. I had a close friend who was in and out of prison in Minneapolis. I knew something of what was happening on the streets in the area. As far as I could see, the police were virtually oblivious to that reality. Further field research in Indianapolis and in Sheffield, England, has confirmed my initial conclusion that trends and patterns in police crime counting tell us nothing about crime itself.

I decided to review the literature on every measure of crime and criminality I could find to see whether any of them, or all of them taken together, would get me any closer to understanding the true nature of our crime problem. In 1980 in *Crime Control Strategies,* I concluded they did not. I could see, however, that the only way for all indices of crime and criminality simultaneously to show that the war on crime was being won was through the decentralization and grassroots democratization of social control.

Leslie Wilkins tells me he doesn't remember, but I clearly recall him telling me that he had sent his advanced research design students on a field exercise. They were to wander through the campus looking for serendipity, because serendipity was what produced breakthroughs in knowledge. It was my serendipity that Paul Jesilow became my next-door colleague. Paul is a student of white-collar crime. When we collaborated on *Myths That Cause Crime,* he gave me a startling insight. For all the crime happening in the streets, there was far more murder and theft taking place in corporate suites. Not only were the police largely oblivious to life in the streets; they weren't even patrolling the places where the great bulk of the crime problem lay. Poor women and children were the ultimate victims of crime, to be sure, but criminally unsafe products, services, and living conditions were killing far more of them than any neighbor was capable of doing with a single gun.

I have also been blessed with another form of serendipity. Earlier experiences, seemingly abandoned for criminology, have come back to help me pinpoint and describe the nature of crime. One of these was my interest in comparative studies and international relations. I had studied a number of languages and had majored in Chinese language and literature. Although I had a strong interest in criminal law in law school and spent two years working as a student public defender, I had an even stronger commitment to the study of Chinese law and international relations. I kept my nose clean, got my security clearance, and worked as a legal intern in the Office of the Assistant Legal Adviser for East Asian Affairs in the U.S. State Department in the summer of 1967. My thoughts of a career in the Foreign Service gave way to despair over the bureaucratic myopia of diplomacy, over the demand for group-think. My third-year law paper won praise from my teacher for its standard Western legal analysis of Chinese Communist diplomatic relations during the first nine months of 1967, when the Cultural Revolution was in full swing. On the other hand, the teacher thought that the concluding chapter—which I loved—offering a Maoist legal analysis was pretty worthless. Another challenge . . .

Now after collaborating with Paul Jesilow, it came to my mind that not only was street crime but a small part of the problem of crime, but crime itself was but a small part of violence. I noticed that U.S. incarceration rates the past century and a half levelled off or dropped only during U.S. engagement in major foreign wars, most recently during the Vietnam War. Our governments seemed to oscillate between building prisons to wage war on killers and thieves at home, and mobilizing and equipping soldiers to kill and steal abroad. People died and suffered just the same regardless of whether the enemy was foreign or was called the criminal element.

This realization brought me to the literature on peace studies, and notably to Birgit Brock-Utne's pioneering 1985 book, *Educating for Peace: A Feminist Perspective.* That book and a gifted feminist sociologist, my wife, Jill Bystydzienski, confirmed my belief that the dynamics of violence at home and in the streets mirrored not only the dynamics of going to war, but the dynamics of a hierarchical social structure that allowed some people to gain wealth, power, position, and legitimation by impoverishing and killing others.

Here serendipity intervened again. In 1961–62, I had followed my parents to Trondheim, Norway, where I finished high school. We were the only Americans in town for most of the year. Late in our sojourn, we were visiting with a Swedish psychologist, Magnus Hedberg. He asked what I wanted to do with my life. I told him I wanted to serve democracy by working for people. "Would you rather work FOR people than WITH people?" he asked. "I don't see the difference," I replied. And he said, "You'll have no trouble going home."

He was wrong. I left home for college five days after we got back to the States, and I'm only now beginning to think I have found a home. Norway was painful for this American adolescent, but transcending that pain and understanding Magnus Hedberg's point implicitly guided much of my research. I returned to give a lecture in Oslo in the spring of 1983, and then went back for the spring of 1986, especially to spend time with Nils Christie and his colleagues, whose work I greatly admire. Oslo is also Birgit Brock-Utne's home. With the help of many Norwegian friends, I was finally to my own satisfaction to describe and distinguish violence from its antithesis. I call that antithesis first "responsiveness," later "democracy." My return to Norway and the insight it gave me are where my intellectual odyssey picks up in this book (chapter 2).

The insight gained from Norway, together with the serendipity of receiving as a present from my wife, Jill, a book on chaos theory, helped me make sense of patterns I had noticed in reviewing the history of crime statistics (chapter 3). The serendipity of being able to help a friend in a political trial further revealed to me how responsiveness or democracy works (chapter 4).

A further important bit of serendipity is the friendship I made with Mark Robarge when we were at Albany together. Mark used to bring windmill designs to my office to discuss. On summer visits of ours, he has a habit of suggesting new things for me to read. About five years ago he lent me Buckminster Fuller's two-volume work, *Synergetics*. Fuller didn't apply his theory to social relations, but I felt intuitively that the link must be there. I tried sketching a few models, then put them aside. It was only after discovering and describing "responsiveness" after the return to Norway, and after applying chaos theory to explaining societal rhythms of violence, that the application of Fuller suddenly hit me. At the risk of overwhelming the reader, I offer a hint here of the ideas and concepts more fully developed in chapter 5.

The beauty of Fuller's geometry is its simplicity. His tetrahedron is the simplest linear three-dimensional way to map any interaction. A measure of learning is our power to generalize from simplicity in an empirically testable way. And by approximation, the theory of the distinction between violence and democracy is testable. Pauline Nichols Pepinsky, a gifted social psychologist and my mother, told me that I was a smart boy, and that if I really worked on it I could make the vague ideas in chapter 2 testable. The tetrahedron presented in chapter 5 transforms "responsiveness" into the specified, testable form I call "democracy."

The other beauty to me is that this geometry provides an explanation of how people become alarmed by violence. It is obvious to all of us that violence represents some kind of social heat, but what form does the heat take? The vital clue to me was the realization that the antithesis of violence has its own form, which I had vaguely described as "responsiveness." I had already postulated that if violence was heat, or entropy, then responsiveness or democracy must be a social coolant, or synergy. Social synergy must have a form of its own.

I had also recognized that violence means being bullheaded, determined to keep heading in the same direction, toward a single goal. So, somehow, social synergy must be the opposite of heading in a straight line.

When I first tried to apply Fuller's tetrahedron to social relations, I modeled vectors of the tetrahedron as dimensions of relations. Now, suddenly, I recalled Fuller's description of how the tetrahedron is composed of two open triangles. Suppose each actor were somehow regarded as one of those two triangles? Slowly, playing at times with pipe cleaners, through successive diagrams of approximation, I found the model offered here unfolding before me.

Finally it hit me that Fuller had described the two triangles as the geometric form for the double helix, the basic structure of living matter. Life itself, as in homeostasis, is synergy. The miracle of life is that so many of us live so long without disintegrating of cancer or by failure of social support. Newton's Second Law can account for cancer, but not for homeostasis. Newton accounts for entropy, and the geometry of the tetrahedron accounts for synergy. Hence the title of Buckminster Fuller's crowning study of the tetrahedron, *Synergetics*.

If alarm over violence indeed increases as people and societies depart from tetrahedronal form, it is easy to understand why. Social relations are themselves perceived as homeostatic life-support mechanisms. Homeostasis is felt throughout the body as tetrahedronal interaction of organ systems. Homeostasis through the social body is felt as tetrahedronal interaction. Violence is instinctively perceived to be a symptom of social illness, of death and decay. Tetrahedronal interaction makes your social body feel healthy and satisfied, rather than hungry and threatened.

The lingering lawyer in me cannot resist considering the practical implications of this theory for an organization I have long been close to, the police. That accounts for the chapter on improving police-community relations through democratic citizen involvement in policing.

My heart, on the other hand, now lies closer to home. As chapter 3 on societal rhythms indicates, no issue is more vital to making peace anywhere, any time, than how we raise and teach our own children. Bob Regoli and John Hewitt have just published a marvelous text on delinquency and juvenile justice, which highlights the children's experience with their own violence and with that of adults.

Jill's and my daughter, Katy, was born in 1977. Never in my life could I imagine a better teacher for me of what childhood is and of what childhood can be than my own daughter. I embarrass her, I know (she tells me so), but I can think of no more authoritative interpreter of my data about childhood and violence than the child who is closest to me. That is the foundation for the chapter on freedom of speech with children. In an era of justified concern over child abuse, that discussion is a particularly fitting way to close this segment of my intellectual odyssey. It brings the book full circle, for I close the next chapter with the prophecy that ageism is the last and most fundamental barrier for humanity to cross on the road to peace. I don't propose to be able to demonstrate an empirical solution to this ultimate bastion of violence, but at least the geometry of violence as departure from tetrahedronality enables me to identify and understand the problem. That's a start. My odyssey will no doubt continue.

I started out wanting to know the nature of crime. I found that crime was a politically arbitrary subset of violence. I followed many hunches, from the back seat of a squad car to Buckminster Fuller's tetrahedron. Some have proved false, others helpful. It has been an exciting journey, which I invite the reader to share.

Many of the sources and ideas in this book will be foreign to social scientists generally, let alone to criminologists. It is not only that social scientists are unaccustomed to drawing on material from fields like modern physics. It is also that I radically reconceive the crime problem, from one of how offenders behave to one of how people's motives interact, from a distinct problem to an inseparable part of the problem of how to make peace instead of war. I have tried to compensate for the strangeness of the overall conception of crime and violence by referring constantly to everyday events that should be familiar to lay readers and social scientists alike. As long as readers of these ideas expect the unexpected, they should have no trouble following my line of thought.

2

Violence as Unresponsiveness[1]

Toward a New Conception of Crime

Emile Durkheim became a pivotal figure in today's criminology by presupposing that punishing criminals is the law-abiding person's blow for virtue. This premise is shared widely from the right to the left by those who believe that young men of the underclass are more violent and more crooked than persons of privilege. Criminologists of the right claim that this criminality is a personal failing of the poor; criminologists of the left claim that it is a mark of extreme social disadvantage. Both sides presume that if truly virtuous and learned persons take control of the state, law will give us order.

Suppose instead that power over others is the major cause of crime and violence: that those who victimize more often and more seriously are those who are less closely supervised, who command greater resources, and who have greater capacity to resist detection and punishment. This is what control theory, opportunity theory, and deterrence theory, taken together, predict. In addition, theories of differential association and identification predict that the more one associates with criminals, the more criminal one is likely to become. Thus if we accepted the logic of these traditional criminological theories, we would expect the power elite in any society to be the prototypic criminal element. This view would mean, for example, that becoming President of the United States would be more conducive to crime than becoming anything except a person with greater power than the President. Of course, power in name may not equal power in fact; empirical studies may even indicate that the President can do less than a Bowery bum. Yet if power is defined as the capacity to subdue others, either Durkheim or the control/opportunity/deterrence/differential association-identification theory must be wrong. By the very logic of control, opportunity, deterrence, and association, those who have the power to punish offenders are likely to be greater offenders than those they punish. It makes perfect criminological sense that CIA and DEA agents move more heroin and cocaine than anyone else (Chambliss 1987), if one only presupposes that domination—power over others—is the fundamental cause of crime and criminality.

What is the relationship of "violence" to "crime" and "punishment"? Durkheim presumes that people punish the crooked in order to stay straight. Yet unless we reject a broad range of conventional criminological theory, it is more plausible to presume that people punish those they can punish to draw attention away from their own crookedness. Punishment—ostensibly designed to return criminality in kind—is inflicted by the more criminal classes on those who have less capacity to commit crime or to resist punishment. The distinctions between crime and punishment are politically partisan—a matter of who defines the situation. The greater capacity to punish contains an index of unilateral power holding, and the standard criminological theories predict that such power holding will increase the propensity to crime. More punishment implies more crime.

There is no direct way of determining empirically who is more crooked. According to many reports, the greatest criminality is found in positions of wealth and power (see Pepinsky and Jesilow 1985). The far greater number of reports to the contrary either may be valid, or (depending on one's presuppositions) may indicate that hidden criminality is greatest where respectability is most apparent. This is a common problem of scientific inquiry; bases of theoretical systems often cannot be subjected to direct testing. No one, for example, has measured whether mass actually reaches the square of the speed of light as it becomes energy. Instead, theoretical systems are compared to determine which system explains the most phenomena with the least complexity. Thus the proponent of a distinct theoretical system must imagine unmeasurable mechanisms as a basis for projecting how observable phenomena might relate to one another.

What follows is an exposition of a theoretical system—how crime and punishment and concentrations of unilateral power or domination might rise and fall if they are only different labels for the same underlying phenomenon. I have begun to explore the explanatory power of this system elsewhere (Pepinsky 1987b); here my objective is not to test it against Durkheim's theory but to describe it, much as Durkheim merely described his explanatory system in *Division of Labor* (1968 [1883]) and later in *Rules of the Sociological Method* (1968 [1885]).

It is instructive to see how explanatory systems first occur to theorists. We have no account of how Durkheim first imagined the functionality of punishment, but in other cases the insight seems to come from letting one's mind escape from the confines of normal inquiry, entering foreign realms of experience, and thus detaching one's mind from the rigors of convention. Einstein is said to have developed his theory of relativity after gazing at a passing streetcar and imagining it to be moving at the speed of light. The transistor was discovered on a whim by analyzing instead of discarding the residue of an experiment that a lab assistant had forgotten to turn off and thus had burned up.

In my case, I became aware of difficulty in describing crime—not to mention the opposite of crime—in the ethos of American political culture, where criminological thought is permeated by the premise that crime is a personal defect that can be subdued only by law enforcers of superior might and virtue. I sought refuge in a political culture where punishment and subordination appeared less acceptable; I wanted particularly to spend time with Norwegian criminologists such as Nils Christie and Thomas Mathiesen, who appear to offer a radical alternative view. I was more comfortable communicating in Norwegian than in any other second language. A Fulbright research grant and supplemental support from Indiana University enabled me to go to Oslo for the spring of 1986 so I could seek to understand how life in a setting of crime and punishment might be distinguished from life in "peaceful societies."

On the surface Norway is much less violent than the United States. Certainly violence is condoned much less widely in public discourse. Ultimately, however, violence simply may be hidden better in Norway than in the United States. While Norwegian discourse has helped me to understand what reducing American levels of violence might entail, it also has made me more skeptical of claims that one people is more or less violent than another.

This is an account of how I came to see crime and punishment as synonymous forms of violence, which altogether rises and falls as systems of power are concentrated and dissipated. The insight came quite unexpectedly as I was translating thoughts about law and social control from English to Norwegian. The result is a theory of violence as "unresponsiveness" and a counterpart theory of peacemaking as a matter of organizing "responsiveness." Here the theoretical system is described merely as a point of departure for testing whether the Durkheimian perspective might warrant being superseded.

If indeed "crime" and "punishment" are arbitrary distinctions for forms of violence, and if indeed the distinctions are politically partisan, then it is morally and epistemologically unacceptable for criminologists to accept any of these distinctions; instead, nonpartisan criminologists ought to develop a theory of violence which presupposes that the only way to reduce the level of crime or punishment in any person or any group is to reduce violence generally. Hence the theory of violence as unresponsiveness is also a theory of crime and of punishment.

Such an initial derivation of a theoretical system is apt to be anecdotal and personal; ultimately the theorist is accounting for a change in his or her own thinking. Even with full awareness that this exposition is unconventional, I feel constrained to present my discovery as it unfolded to me. I am in the frustrating position of believing that the nature of crime and punishment has become far clearer to me, while finding at the same time that conventional proof of "facts" about crime and punishment has become more elusive. My questions have become clearer as my ignorance has become more manifest. If the history of science is any guide, under the best of circumstances only considerable time and as yet unimaginable exploration will tell whether my new-found confidence and skepticism are misplaced.

THE NORWEGIAN CONTEXT

I went to Norway in the spring of 1986 in hopes of finding relatively nonviolent people and of learning how they experience human relationships. I came from the American heartland, where I saw violence constantly; worse, I lived among people for whom even "help" generally took the form of a repressive response to problems of human relationships. I felt that my pursuit of criminology suffered in so violent a climate. I was not even sure I could take myself seriously if I did not start with the premise that offenders need to be hurt both to teach the offender a lesson and to demonstrate sympathy for the victim. I felt stilted, unreal, and overabstract when proposing nonviolent responses to crime (Pepinsky and Jesilow 1985; see also Tifft and Sullivan 1980; Quinney 1987, 1982). I was intrigued to learn that nonviolence was prominent in Norwegian criminology (Christie 1981; Mathiesen 1986) and in the related literature (Galtung 1969). I knew that on any given day in Norway one person in 2,000 was in jail or prison, while in the U.S. about one American in 250 was incarcerated. I also know that Norwegian police figures for violent crime were dramatically lower than their American counterparts. In Norway perhaps nonviolence would be palpable and workable.

In one respect I think I was right to suppose both that Norway is nonviolent and that in Norway I could learn more clearly what nonviolence entails. In another respect I am more confused than ever about distinguishing levels of violence, and I have more difficulty establishing to my own satisfaction that Norway is nonviolent. In this discussion I will attempt to share both the clarification and the confusion, which together suggest an intriguing way to address and study crime and violence.

Briefly, Norwegians indeed are formally less inclined than Americans to accept violence. Although twelve Germans and twenty-five Norwegians were hanged as war criminals just after World War II, I can scarcely imagine any Norwegian today conceding that capital punishment has redeeming value. From time to time I was asked, almost in astonishment, whether some Americans truly celebrated executions or simply favored them on the grounds that some murderers do not deserve to be fed and housed in prison. In foreign policy, Norway, a NATO ally, meets Sandinista leaders and gives them nonmilitary aid.

Consider the terms of debate at the January 1986 meeting of KROM, the Norwegian Organization for Criminal Reform. Some of the most notorious offenders in Norway, including those convicted of bombing and stabbing deaths and of armed robbery while off-duty from police work, were furloughed from a maximum-security prison to go into Oslo unsupervised and to catch buses to a mountain resort. The warden and other officials and prison workers met with them to hear and respond to inmates' grievances. The prisoners conceded that there was virtually no physical threat in the prison other than occasional strip searches, while the officials acknowledged that, by law, prisoners deserved no punishment beyond being forbidden to leave prison without permission (see Falchenberg, Letvik, and Snare 1986).

Aided by staff members from the Department of Criminology and Criminal Law at the University of Oslo, guards and inmates at another prison held a joint press conference in June 1986. At that time they announced to prison authorities the results of a study indicating that more stringent efforts to find and curtail drugs among inmates would be counterproductive (Letvik 1986).

Although there was talk about tougher prosecution of men who beat or raped the women they lived with, even the committed feminists I talked with, such as those in the Department of Women's Law at the University of Oslo, were equally adamant that violent treatment of the men represented surrender rather than opposition to the male definition of the situation. Time and again in what appeared to be routine discourse, Norwegians avoided supporting the violent, repressive countermeasures commonly advocated in the United States, although they condemned the original violence. Norwegian constructs of human events are remarkably, cogently nonviolent.

The confusing element is trying to understand how violently Norwegians actually behave in comparison to Americans. Do Norwegians talk more peacefully than they act? I was warned repeatedly that peace in Norway was more apparent than real, and I was told that the impression of peace dissipated after an immigrant or visitor had spent some time among Norwegians. The Norwegians reported that they were only beginning to recognize a great deal of previously hidden violence, as in the home, in neighborhood boundary disputes, in schools, or against minorities such as gypsies or Jews or darker-skinned immigrants (see, e.g., Johansen 1984). Occasionally I saw violence in situations where a Norwegian would argue that there was none. I was cautioned not to allege violence (as in hospitals; see Pepinsky and Jesilow 1985) where "scientifically" speaking, the existence of violence was open to honest dispute. More than ever I learned to doubt and distrust reports of how much violence has or has not occurred in any setting. As a result I find myself incapable of demonstrating that violent behavior is more prevalent in the United States than in Norway. In this respect I know less about violence than I once thought I knew.

RESPONSIVENESS

My understanding of a nonviolent construct of the world came from translating my previous thinking into Norwegian for a series of invited lectures. I wanted to describe the notion that Jesilow (1982) had introduced into our work (Pepinsky and Jesilow 1985), namely that American law and government favor larger corporations, where the law limits the liability of investors. Adam Smith (1937 [1776]) pinpointed the difficulty to incorporation: Limiting the liability of corporate owners to what they invested in the corporation made the risk of investing with strangers acceptable; this in turn permitted corporations to become large enough and reckless enough to monopolize markets and to ignore the welfare of consumers.

"Liability" translates into Norwegian as *ansvar,* usually translated back into English as "responsibility." While working in Norwegian, I found myself in effect contrasting what legal practice does most often and most secretly to young men of the underclass by "holding them responsible" for social disorder with the corresponding legal practice of "limiting responsibility" of the corporate elite. The act of incorporation itself limits responsibility; thereafter, those most likely to be excused for social disorder, such as corporations, are excused on the grounds that they do too much good to be sanctioned seriously for offenses. The larger corporations are those most likely to be bailed out by the state and to receive tax breaks and major military contracts. American Governments characteristically limit the responsibility of more powerful people for social disorder; when disorder threatens, U.S. governments, from local to national levels, propose to restore order by holding weaker persons responsible. Legal recourse against the powerless for the disorder that occurs when the powerful are relieved of responsibility amounts to political sleight of hand.

In a subsequent lecture I outlined a proposal for increasing police accountability to residents of patrol districts (Pepinsky 1984). In Norwegian I proposed to make policy *ansvarlig,* the term used by Norwegians for making offenders "responsible" for crimes. In Norwegian thinking, "holding responsible" is the same as "making accountable" and "imposing liability."

"Responsibility/responsible," the standard translation of *ansvar/ansvarlig,* is not literally correct. It is more accurate to translate *ansvar* as "responsiveness."[2] As nearly as the Norwegian concept can be rendered into English, criminal penalties are imposed to "make offenders responsive." Proposals like mine would also "make police responsive," while the problem with creating and favoring corporations is that incorporation "limits responsiveness."

As I see it, *"responsiveness" means that what one expects to achieve by one's actions (in common law parlance, one's "in-*

tent") is modified continually to accommodate the experience and feelings of those affected by one's actions. "Responsiveness" means doing things *with* people rather than *to* or *for* people.

The issue of police accountability or "responsiveness" is only one aspect of a larger issue: whether politicians and officials become responsive to the chaos that ensues from condoning and subsidizing the unresponsiveness of the corporate elite. Being responsive to disorder entails trying to create order by regarding oneself as a part of the problem. Responsive persons or groups work on problems by changing themselves. "Responsiveness," as the Norwegian translation suggests, means "holding oneself liable" for problem solving.

It is understandable that politicans, officials, or any other people who aim to hold their jobs and their status would be reluctant to acknowledge personal liability for a chaotic economic order. The politician who says, "I'm sorry so many people are hurt by economic turmoil, and I accept personal responsibility for improving the situation" is likely to be driven from office when problems remain. Jimmy Carter lost his presidency in this way. How much safer it is to pass the buck by inviting those on whom one's incumbency depends most strongly to join in attributing chaos to others—ideally to people too weak to fight back. As Brogden and Brogden (1982) show in the case of Great Britain, politicians have achieved widespread interclass consensus as to the legitimacy of the ruling structure by attributing domestic chaos to young men of the underclass, our prototypic "street offenders" or "criminal element." Purporting to make underclass young men responsive in regard to disorder is a way of letting officials and members of the elite off the hook for being responsive themselves.

Although one can speak in Norwegian of making offenders responsive to crimes, the idea is seldom mentioned even by defenders of punishment like Andenaes (1966), who stress "general prevention." This expression means getting people to do what the law demands before they have to experience the punishment. Indeed, if general prevention is achieved, there should be scarcely any call for punishment (Pepinsky 1978). Thus a Norwegian proponent of punishment, as for drunk driving, would be more likely to cite a high degree of conformity to law as evidence that punishment works than to call for more incarceration. This observation highlights the absurdity of the fondness for "holding offenders responsible." "Holding" responsible entails depriving offenders of liberty—keeping them in a fixed location, determining what they do, making them passive vessels for our own actions rather than increasing their opportunity to respond to what people do to them. Officials who limit the responsiveness of power elites, and then limit their own responsiveness by passing the blame for disorder to underclass young men, ultimately reduce even the capacity of "the criminal element" to be responsive by punishing them. In this system of law, unresponsiveness feeds and builds on itself.

In my youth, American schoolboys in cafeteria lunch lines played a game in which we punched the boy behind us on the shoulder as hard as we could and said, "Pass it on!" The law-and-order politics so familiar to Americans is essentially a game of "pass it on." Rather than responding to disorder by trying to restructure situations with people at least as powerful as ourselves, we tend to descend into violence by passing disorder on to persons weaker than ourselves and less able to pass it back. Some people can hire others to pass on violence and disorder to large groups of victims, but young men of the underclass may have only one another or women and children to abuse. The saddest aspect of this game is that disorder and violence have nowhere to go but up, as everyone "passes it on" downward.

Unresponsiveness at any level is the element that makes people call action "violent."[3] People harm one another in many ways without their actions being termed violent. We have other labels for such actions, such as "accidental" or "misguided." Violence entails a willful disregard for one's effect on others. That disregard in and of itself may be sufficient for us to classify an action as violent, as when we discuss child "neglect." Violence or disregard for others may be direct and personal or indirect and structural, as when the plight of impoverished classes remains unaddressed by privileged classes in a shared economic order. Individual disregard for others is mirrored by organized disregard at higher levels of the social system. From the personal to the structural level, unresponsiveness and violence are essentially the same.

The question becomes, "How does violence as unresponsiveness grow or dwindle?" The passing on of violence or unresponsiveness through law suggests that violence grows by reproduction—that unresponsiveness begets unresponsiveness, which may or may not happen to be called "crime," depending on the political circumstances. Violence, including crime, can be abated only by confronting it with its opposing force—responsiveness. Let us begin by exploring the dynamics of unresponsiveness.

VIOLENCE AS UNRESPONSIVENESS

The first distinction drawn by Christie (1981) between violent and nonviolent communities[4] is that in many respects nonviolent people know one another as individuals. Imagine yourself a witness offering to testify at two separate sentencing hearings for someone convicted of taking a television set out of an empty house. In one instance you are the victim, who has never met the defendant. In another, the defendant is your brother or sister. In the latter case you are likely to find more redeeming features in the defendant to mention to the judge. More to the point, probably you will want to generate greater sympathy and lenience toward the defendant. You will feel more of the defendant's pain and will decide more rapidly that the defendant has suffered enough. In addition, you will more likely recall good and beneficial things that the defendant has done for you and for others "when given the chance." In a word, you will be more responsive to the defendant in the latter case, and therefore less inclined to violence.

Obviously, people who live close together can treat one another with hideous personal disregard, as in domestic violence. Although some form of physical or spiritual proximity is necessary for intimacy and nonviolence, proximity alone is not sufficient. People can become so brutalized that they give up on intimacy altogether, or they can be victimized so heavily by forces they feel powerless to resist that they lash out blindly at persons who are harmless because they are weak or forgiving, perhaps even because they are loving. At this juncture the tragic, self-defeating nature of violence becomes most apparent.

Violence results from depersonalization, which can be induced in any number of ways, as by putting people in the bureaucratic position of doing things to others without responding to their experience and their feelings. Zimbardo (1978) achieved this result by introducing some students to others as prisoners. Milgram (1973) achieved it by putting victims behind glass and having an experimenter turn electric shock into a mechanical exercise. Denzin (1982) explores how a murderer detaches himself from the victim, much as people who have dropped bombs in war describe how the human beings below disappear into insignificance. A broad array of literature indicates how stereotyping leads to violence (Pepinsky 1977). At the other end of the scale, allowing the victim and the offender to meet and talk about how they feel and what they want can lead to surprising displays of nonviolence, as when a police officer forgives an assailant (Witcher 1986).

So far, the obvious. It is tautologous to say that one cannot continue to hurt another person while feeling and responding to that person's pain. Because violence means forcing on others what they do not want, responsiveness is antithetical to violence. It is almost as obvious that one is less likely to start hurting a person one already knows and treats with compassion. After all, the best predictor is that what has been will continue to be.[5]

Now to the less obvious. Some of the subjects in Milgram's experiment refused to give the victims stronger shocks. That is, some people are less inclined than others to hurt strangers. If responsiveness makes the difference between violence and nonviolence in these cases, it must be that the experience of responsiveness or violence in one setting or with some people carries over to other persons and other settings. This transfer is supposed to occur often, as in the proposal that abused children are prone to become child abusers, or that the man who is browbeaten at work by the boss goes home and beats his wife, or that an Islamic child whose parents are killed by an Israeli bomb may grow up to shoot innocent Christians in East Beirut.

The theory is that violence begets violence, whether the offender has committed or received unresponsiveness in the past. Giving and receiving unresponsiveness produce the same result; the master fears losing his or her grip on the unwilling subject, while the subject seeks a way out. An authoritarian leader, no matter how wealthy, fears the prospect of falling from power as deeply as the subject fears being annihilated. In either case, unreciprocated responsiveness creates the potential for violence. Those who become habituated to following orders and others who become habituated to giving orders are equally liable to violence in other situations.

Unresponsiveness can become organized. It may inhere in a particular tradition or fiat in which a husband, father, boss, or officer is always right; it escalates when broader kinship, professional, or state networks unite to punish insubordination whether or not the commands should have been followed. The voltage stored in the social system rises when a court supports someone's right to decide someone else's fate confidentially; when it fails to order someone to explain what is done to someone else; when it fails to listen to a complainant asking to be heard; or when a power broker refuses to negotiate with opponents. Firm intergenerational class lines and enduring power imbalances of gender or age heighten social tension. The higher the geocentric concentration of intergenerational class division, the higher the voltage, and the more highly charged and concentrated the discharges of violent power. The United States and the Soviet Union are dominant centers of violent activity in today's world. They lead in all indices of outbreaks of violence. Americans and Russians excuse much of the repression in their home countries on the grounds that the other country poses the greatest threat to peace and that their own nation offers the greatest hope for peace. Some peoples, such as Nicaraguans and Poles, believe that the closer center of violence is the devil, while the more distant center offers some hope of escape from bondage. Others—Muslims and Buddhists among them—see that the Soviet-American axis is shifting toward an alliance against the more southern, darker-skinned nations. The more thoroughly divided geopolitically the world becomes, the greater the threat of human extinction.

Thus violence and responsiveness operate by the same principles at all levels, from the interpersonal to the international. Every human being lives at all of these levels simultaneously, and is at once the subject and the object of both violent and responsive energy. Crosscurrents of violence and responsiveness run constantly in all of us, and help to account for perversity and unanticipated behavior at any given level.

If the analogy to building up electric tension holds true, trying to predict outbreaks of violence is like trying to predict where lightning will strike. Some towers are struck constantly in storms; and, in like fashion, it appears that poorer and less powerful people are hurt with the greatest frequency, most often in regions where unresponsiveness is concentrated most heavily. Lightning strikes young American black men, for instance, with remarkable rapidity in such forms as imprisonment, shooting death, and drug poisoning. Yet violence can be predicted only as probabilities, which only sporadically become extensive in a single place such as Kampuchea or Ethiopia. Violence occurs in human interaction; it is generated by the concentration of unresponsiveness in the area and by the availability of victims too weak to resist. The higher one's position of power, the less restrained and the more damaging one's violence becomes. Even so, blame for violence cannot be apportioned. Offenders are infected as deeply as victims with the disease of violence; offenders in one setting certainly are victims in others; the existence of one victim implies the existence of other victims.

It is fair to suppose that status and wealth make one's offenses more frequent and more damaging, but not because one becomes a worse offender. Whoever sold the TOW missiles illegally to Iran committed more cold-blooded and more numerous murders than did John Wayne Gacy or Charles Manson, but not because politicians are more depraved than street offenders. Meanwhile, at the level at which we usually try to measure, violence turns out to be essentially unpredictable (Monahan 1978). To paraphrase Schumacher's (1975) maxim, individual acts of violence in principle are unpredictable.

NONPARTISAN CRIME THEORY

So far I have offered a theory of violence that mentions crime. The question remains: Should crime be explained in its own right as a distinct form of violence? I believe not. One way to illustrate this point is with the following description of an event:

Q: What was the harm?
A. Someone died.
Q. What was the proximate cause of death?
A. A knife cut an aorta.
Q. Had the person who died given permission to be cut with the knife?
A. Apparently not.
Q. Did the person who died physically threaten the wielder of the knife?
A. Apparently not.
Q. Did the wielder of the knife want to cause death?
A. It appears not, but the wielder apparently did intend the cutting and had reason to know that it posed a serious life risk.

Here is a description of an act and its consequences. The description contains some subjective inferences about permission, provocation, and intent. Yet the statement avoids clues about setting and other details that might indicate the status of the parties. Is a crime stated? Observers will differ, but consensus probably can be approached by describing *who* the deceased and the knife wielder are. If they are a young couple or a pair of customers in a bar, it may quickly become "obvious" that an offense, manslaughter at least, occurred.

Suppose the knife wielder and the deceased are surgeon and patient; suppose further, as reportedly occurs in a substantial percentage of Norwegian surgical procedures (*Aftenposten* 1986), that the patient had not understood what the operation was, why it was needed, or what risks it entailed. In

American law a person who is not informed about an operation cannot consent legally to the operation. In addition, operating without permission is criminal assault, is it not? Drawing on Pepinsky and Jesilow (1985), I have suggested in many lectures that surgeons in such cases are guilty of manslaughter or murder. Yet people in the audience always insist that the surgeons are certainly innocent.

Even if one accepts the arbitrary political distinctions of a particular penal code, crimes cannot be specified without using terms that connote something about the status and the relationship of the parties involved. If the scenario had said, for instance, that the victim had been "stabbed" or "attacked" rather than "cut," wrongful intent might have been implied. By saying, however, that a surgeon "operates on" rather than "stabs" or "attacks" patients, we lower the odds that surgery will be perceived as criminal. These implicit cues as to *who* is involved in an act permeate events from which crime indices are constructed. Crime cannot be specified without at least such implicit references. Among the stronger biases involved in specifying crime, persons of wealth and position are "not really criminals," so that the things they do, such as surgery, are extraordinarily unlikely to be perceived as crimes.

On the other hand, this vignette captures the similarity between the operation and the barroom fight. In neither case did the wielder of the knife obtain permission to use it from the person cut. One can suppose that the surgeon meant well while the fighter meant ill, but the surgeon presumed that the patient had nothing relevant to say about the desirability of the operation as surely as the fighter ignored the sensibilities of the person stabbed. This attitude—that the person affected by one's action has nothing worthwhile to say in the matter—is the essence of unresponsiveness. This personal disregard allows doctors to decide that they know what needs to be done without consulting their patients; the same mechanism allows muggers or rapists to accost their victims, bombardiers to release their loads, despots to rule their people, or beneficiaries to accept class privileges. In each case it is likely that those who display the disregard have learned to do so by being the subjects of unresponsiveness themselves.

Occasionally it is recognized that crimes are not simply

acts, but that they entail independent reactive judgment based on the perceived status of the parties involved, or on beliefs about the appropriateness of using penal sanctions (see Christie 1981; Hulsman 1986; Wilkins 1984). "Crime" is part of a much larger universe of events that could be classified similarly but are not. Even without changing a penal code, we can infer equally plausibly on the basis of social data that poor people or rich people are by far the worst offenders. Sometimes, as in China just after 1949, political change can cause dramatic redefinitions of who is criminal—peasants one day and landlords the next.

Should criminologists build one body of theory to account for acts when the actors are perceived to be criminal and another body of theory to explain the same acts when the actors are perceived to be law-abiding? This is the common practice in criminology. One might suppose, for instance, that someone who wields a knife in a bar acts out of disregard for human life, while the surgeon wields the knife to save or enhance life. I am inclined to reject such distinctions for several reasons.

First, I draw from Norwegian experience. As I stated earlier, Norwegians on balance appear to be much less violent than Americans. In their rhetoric, they are relatively loath to recognize the general inferiority of any class of persons within a community. Whereas Americans appear to assume that people have an equal chance to succeed and have no one but themselves to blame for failure, Norwegians believe that people should be regarded as equals in social status, although there are signs of a "conservative" tilt in the American direction. I have no way of knowing whether Norwegians are really less violent than Americans or whether their rhetoric reflects their true beliefs, but the logic of surface appearances is compelling. The acceptance of class distinctions is a sign of violence.[6] Learning to explain crime and violence without drawing invidious class distinctions ought to give people greater power to live without violence.

Second, theories that criminals are poor or deprived imply the premise that possessors of power and privilege have earned their true reward; they and their actions are assumed to be the safest and the most valued in the community. The pervasiveness of this premise in criminological work explains why so many criminologists are accused of being servants of the state; this is partisan, political criminology of a kind that Weber (1946 [1918]) rightly rejected. Scholarly independence requires that a theory of crime describe and explain behavior free from influences of class and power. As Nils Christie pointed out to me, we are thus required also to assume that the rich and the powerful are no more violent or crooked than the poor and the weak. Whatever theory applies to the surgeon must also apply equally and in the same way to bar patrons who wield knives; otherwise it is a political position that supports structural violence (Galtung 1969). It is simply too convenient to presume that doctors who wield knives are defending life while bar patrons who wield knives oppose life—too convenient because these distinctions almost invariably give people in privileged positions the benefit of the doubt.

Third, acceptance of prevailing biases as to who is more violent and more crooked becomes theoretically absurd in cross-cultural studies. Because different kinds of people achieve power for different reasons in different political systems, there can be no general theory beyond birthright to explain how one becomes a surgeon rather than a bum in any society. Hence the cross-cultural theory that the mighty are more virtuous than the weak has no substance.

Finally, the law of parsimony is a sensible guide to theory building. Start with the simplest way of explaining the most phenomena; build on it by introducing the fewest possible qualifications to account for empirical disconfirmation. If a single explanation accounts for any act that *could* be considered a crime, such an explanation is preferable to disparate theories that account only for subsets of this action.

A decade ago I proposed that "appropriation"—depriving others of *future* use of a resource—included all acts that conceivably could be called crimes (Pepinsky 1976a:25–43). Elsewhere I proposed that crime as defined by any of the standard measures increased with geographic mobility (Pepinsky 1975b, 1980). Yet I could not pinpoint why appropriation should encapsulate crime, nor could I specify how appropriation related to mobility. Now I perceive in "unresponsiveness" an underlying cause for violence and for fear of crime.

In view of human reliance on learning for survival, one can assume that an alarm sounds when people perceive that others on whom they depend are unresponsive to them, or that they themselves are unresponsive to the others. Alarm mounts as the unresponsiveness persists and affects the supplying of material needs. To relieve alarm concerning unresponsiveness, systems may be designed to make people dependable. One such system is known as "bureaucracy," according to which people are believed to become reliable and predictable if they are organized to do what one wants wherever and whenever one determines. The effect, however, is to impair responsiveness still further. No set of substantive rules can determine its own application in unforeseeable contingencies (Pepinsky 1984). People ordered to enforce such rules are reduced to devising ways to conceal their unavoidable exercise of discretion (as can happen in policing; see Pepinsky 1975a, 1976b). Under these circumstances managers and subordinates alike sense a growing failure to communicate. Often, in a vicious circle, they try to monitor and specify rules still further. Umbrella systems are created in an attempt to bring together and to close open, interdependent systems. If the process continues, five-year plans and corporate structures can determine the fate of hundreds of millions of people at once, making the subjects feel that they have no part in shaping their own destinies and making the task of presiding over the organization seem totally unmanageable. Leaders and subjects are increasingly inclined to cut off others still further, and the result is mounting violence. The more they experience violence, the more insensitive to violence they become.

People often try to relieve their alarm, in which case they may give it human form and then try to exorcise it. If one has little enough to lose and perceives some chance of success, the human devil may take the form of one's superior, from parent to president to ruling class. Many people, however, cannot imagine living outside the prevailing order, so they try instead to achieve consensus on defining and describing enemies too weak or too distant to threaten their own position or to be confused with themselves. In the U.S. today, the most widely accepted enemies are foreigners called Communists and native underclass young men called the criminal element. We project onto them the unresponsiveness we feel; we construct images of the unresponsiveness we experience as criminal or enemy exploits; and we vent our rage and alarm at the enemies paraded before us in courtrooms, in death houses, and on battlefields.

The prototypic offender has no concern at all for how his behavior affects others, and does not allow others to affect him. Vices like drug use can be described as embodiments of global unresponsiveness. Someone you trust may take something from you without asking and perhaps may cause anger, but will not cause you to call the police. Yet if a stranger breaks into your house, even without taking anything, your fantasies of unresponsive people killing you in your sleep can cause fear and recrimination. The break-in creates even greater alarm among persons who already feel isolated or inundated with unresponsiveness; thus members of certain groups—such as U.S. residents or elderly city residents living alone—may fear one particular "crime" more than others (Merry 1981). Crime is a real source of anxiety, but less so than the most celebrated acts of unilateral (unresponsive) action, such as a presidential directive. The anxiety over the totality of unresponsiveness is channeled into fear and loathing of criminals.

ORGANIZING RESPONSIVENESS

If (as proposed) violence builds when people sense that their fate is determined unilaterally—either by others or by themselves—then the alarm must be relieved, and the force or the voltage of violence must be reduced, by conveying that those affected by action share in creating the action. The force of violence can be relieved only by giving people the sense that they influence the events that shape their destiny even though they do not determine those events.

The primary qualification for influencing decisions is to determine how directly one is affected by the outcome.

Worker and customer together are better qualified to decide how to produce than the absentee owner or the academic consultant (a position well represented today in the U.S. by the magazines *Changing Work* and *Building Economic Alternatives*). As recognized in *Roe v. Wade,* a pregnant woman is best qualified to decide whether to continue a pregnancy. Victim and offender together are better qualified than judges to decide how to respond to crimes (Christie 1977). Democratic authority is based on this belief: that the best decisions are made by those whose fate is on the line. In contrast, all the forms of authority described by Max Weber entail the belief that one person is qualified to make decisions for others.

"Responsiveness" recognizes that the action of anyone who lives among people affects others asymptotically from face-to-face interaction and owner/subject relations to the outer edges and the far-removed pockets of humanity. On the one hand this idea means that even transnational government is requisite to peacemaking, provided that the power of governments becomes successively stronger as one moves "downward" to the level of face-to-face interaction among those who live and work together. On the other hand, even at the strongest, most basic local level of government, responsiveness requires that every person's input be crucial to decisions, but without determining them. That is, the decision would be different if the input of any affected person or group were withdrawn, but it could not be predicted from the input of any single person or group, just as the value of an interaction term cannot be predicted from the value of any single constituent variable.[7]

Therefore, organizing responsiveness takes the form of decentralizing and democratizing institutions (Wilkins 1984), which seems to reduce fear and loathing of crime (Pepinsky 1986). Norway appears to have reduced incarceration by 70 percent in the latter half of the nineteenth century because political power devolved on localities (Pepinsky 1987b); now it is supposed that violence may decline further as women's participation is expanded to constitute 40 percent of the government (see Brock-Utne 1985; Stang Dahl 1985, 1987; for American thought see Harris 1985).

Varied expressions of responsiveness occur routinely in the

United States, which is large, diverse, and experimental enough so that they can take place as though underground. At the transnational level in the Witness for Peace program, Americans have taken the unprecedented step of representing a military power by traveling thousands of miles to Nicaragua to put themselves in the way of their own government's weapons. At the local level even entire cities are placing the control of urban economies back into local hands. As Mayor George Latimer (1986 [1983]) says about Saint Paul, "The Homegrown Economy presents a new orientation for the city and a new approach to using our community's resources . . . nurturing economic growth from within, instead of waiting on outside forces."

Once a criminologist forsakes the premise that some community members are more law-abiding than others in favor of the premise that crime and violence are a potential shared by all community members—which can be relieved by reducing the scale of unresponsiveness—crime prevention becomes not only imaginable but practical. Indeed, crime prevention is part of the everyday experience even of people as outwardly violent as Americans, who have ample peacemaking traditions on which to build.

The fact that our species has survived on this planet for more than a few generations is a tribute to human responsiveness. We breed far more slowly than other species and tolerate a much narrower range of environmental variation than, for example, the cockroach. The survival of our species rests on the prolonged, extraordinary nurture of species members from gestation to maturity, on exchange of information, and on learning complex and radical adaptation to a rather capricious environment. Our survival for some hundred thousand generations must mean that cooperation, nurture, compassion, and responsiveness are the overwhelming reality of human existence. The corollary fact—that so many of us notice only that "human nature" entails violence, competition, and paying one's own way—must mean that we are highly sensitive to our own unresponsiveness, not that violence and competition are a dominant characteristic. We are so sensitive to this threat that the sound of our "unresponsiveness alarm" can drown out our awareness of our own compassion. Human nature is both cooperative and competitive, both responsive and violent. Learning to make peace in place of violence entails not a change in human nature but a recognition that our nature has a compassionate side, as well as recognizing our ability to rechannel energy and investment from our violent side to our peaceful side.

CONCLUSION

After years of struggling to dig beneath political artifact, I think that at least I have here begun to understand what distinguishes violent and criminal action from nonviolent and noncriminal action. I can see this distinctive phenomenon—unresponsiveness—reproducing itself as it moves through and around social orders. The regeneration and accumulation of unresponsiveness appear to follow some of the basic principles of electrophysics; these principles explain why crime remains intractable and even appears to grow worse in the face of concerted attempts to discipline offenders and would-be offenders.

Perhaps it is even more important to begin to understand the nature of the relationship between crime and violence. We have pursued the study of crime and punishment, which are indistinguishable and mutually reinforcing when political identities are removed. Peacemaking action, however, has a distinctive character, not in the behavior of a single actor but in how people interact. Responsiveness can be described as a personal pattern or as a form of organization which can be called "democracy." Here is a mechanism by which violence can be relieved at any social level; here is a true possibility for crime prevention.

I take it as axiomatic that most of the contemporary thinking about crime is true: that opportunity, failure of control or deterrence, and association, taken together, account largely for criminality. I have not tried to document the support for these categories of theory; if together they are found false, the implied theory of violence as unresponsiveness fails too. The theories are so well established and so voluminously documented, however, that I see no point in defending them here.

Most of the propositions in the theory of violence as unresponsiveness are reached deductively in a series of syllogisms. They are set forth because they follow from established explanations of crime and from well-known patterns of punishment. In this regard the theory is most obviously subject to testing; it must be revised or rejected if and when any one of these propositions is found not to conform to empirical reality.

The other major constraint on this theory is parsimony. The key for me lay in considering how three legal phenomena in the English language became one when I translated my thoughts into Norwegian. Insofar as the propositions in the theory remain intact, the dynamics of responsiveness/unresponsiveness account for the gamut of human violence and subsume a number of criminological theories which have been presumed until now to be discrete, at best an array of addictive factors in multicausal models of criminality.

By definition, empirical tests of theories of crime or crime control, violence or peacemaking rest ultimately on comparing quantities of crime or violence in samples of human behavior. Although mistakes can be made in "finding" crime or violence where in reality none exists, in many cases we can be confident that crime or violence actually has occurred. We can see people being shot on city streets and on battlefields; we can be fairly confident of at least some prison censuses; and many instances of crime and violence can be counted.

The problem is that crime or violence works as unresponsiveness is supposed to work; that the crime and violence we count are bound to be infinitesimal relative to the hidden crime or violence; and that levels of the one may be uncorrelated to levels of the other (Pepinsky 1987a). Varying counts of crime or violence may correlate highly; for instance, police-recorded burglary and victim-survey burglary may correlate closely with the status of the complainants. Yet as I said above in the discussion of nonpartisan crime theory, this phenomenon may well mean only that the biases of recorders and of complainants or respondents in these cases are fairly constant but nonetheless overwhelming and problematic.

This theory of violence as unresponsiveness, and the research from which it is drawn, hold that one bias is fundamental: Perpetrators, victims, and observers of underclass crime and violence are much more likely to recognize and perceive crime and violence than are perpetrators, victims, and observers of elite crime and violence. Indeed, theory and research suggest that "respectability," "achievement," and "status" themselves are essentially masks for crime and violence; if standard theories of control, opportunity, and deterrence hold true, crimes committed behind these masks must cost people far more lives, liberty, and property than the transgressions of the poor. Crime is normal because all of us commit so much of it.

Even among the poor, as we discover to our sorrow, hidden violence and predation, as of women and children in the privacy of the home, may be far more prevalent than we ever supposed. Lower rates of domestic violence may simply reflect greater concealment and denial of crime.

Some people would claim that failing to notice crime and violence is the greatest crime and violence of all. On the other hand, both silence—which provides a mask of respectability—and approval are also consistent with responsiveness and peace. Thus in the space left unexplored by our research methods, where our counts of crime or violence end, violence or responsiveness may exist in any relative amounts. Scandinavians, for instance, have warned me repeatedly that Norway is far more violent than I can imagine, and that it is fallacious to consider Norwegian society more peaceful than American society.

Thus when crime and violence seem to have increased, they may actually have been reduced (as some child abuse and domestic violence experts claim), because our findings rest on having unmasked hidden crime and violence. Conversely, when we find that crime and violence have decreased, as people have told me occurred in Portugal under Salazar's dictatorship, crime and violence may in fact have become more protected and entrenched, more hidden and widespread. These possibilities apply even to honest personal reporting. How many despots believe in their own virtue? How many of us are subject to similar personal blindness? Not even a polygraph can tell.

All of this means that in all candor I can imagine no way to test whether responsiveness and unresponsiveness tend to reproduce themselves. I comfort myself with the belief that all other criminological theory on this matter is as speculative as mine.

It also strikes me that growing violence is entirely consistent with growing responsiveness. Depending on one's viewpoint, the level of global violence emanating from the U.S. is at a historical high, but so is the level of peacemaking. The development of information and transport technology has facilitated the spread of both war and peacemaking, both violence and responsiveness, both crime and honesty. If our species, by some continuing miracle, should survive for further millennia, we can expect confrontation and cooperation to expand side by side to a considerable degree. Not only are the East-West axes in the North giving way to a North-South axis of confrontation, sexism is now being noted globally; and if women ever come to share power fully with men, we may achieve unity between the genders on ageism—the confrontation between middle age and the extremes of youth and maturity, in which marginal children and elderly people are played off against one another on a global scale. At the same time, as in the Witness for Peace program and in many quieter ways, people are traveling, mingling, communicating, and responding to one another with unprecedented global compassion. The theory of violence as unresponsiveness predicts that these contradictions between mounting violence and mounting responsiveness exist not only internationally but also in the soul and the actions of every single human being.

This is the long-range, millennial view. Within single lifetimes we experience the progress of responsiveness and violence differently. When unresponsiveness is aggregated, as in the U.S., it follows a cyclical pattern, and each of us normally passes through two or three contrasting cycles in a lifetime. It is a mistake to take any of these cycles as indicative of an enduring trend. It is a challenge to understand how the cyclical pattern can be broken so that a society emerges peaceful. This cyclical pattern is the subject of chapter 3.

3

Societal Rhythms in the Chaos of Violence

THE REMARKABLE SIXTIES

It is well known among criminologists in the U.S. that police-recorded crime began climbing in the latter 1950s, and was exploding by the latter 1960s (see for example Selke and Pepinsky 1982). The same occurred in Britain (Pepinsky 1987a) and in Norway (Christie 1982). In each case the crime wave was seen as a problem of young people getting out of hand. Each crime wave was explained in terms of conditions thought peculiar to its own society. Foreign influence and mobility were undermining Norwegian solidarity. British youth did not know the privation of war. Children in the U.S. had known only prosperity.

By the time the crime waves were in full swing, youthful unrest was in many ways apparent. From an American vantage point, it seemed as though the unrest was a reaction to peculiarly American hypocrisy and state violence (Klein 1969). Where we saw unrest occurring elsewhere, as in France, we were inclined to think that our own protest movement, or the insidious influence of rock and roll perhaps, had infected other countries.

Several years ago I had the privilege of getting acquainted with a Chinese sociologist who had been punished and disciplined repeatedly for his bourgeois ways, in the fifties, sixties, and seventies. The late Great Proletarian Cultural Revolution, kicked off by university students in 1966, was much on his mind. He smiled as he observed that youthful unrest in China had had great international impact. Why, it had even inspired riots and mass protests by young people in the U.S. in the late sixties. And of course this unrest was traceable to the Gang of Four's evil influence, supported by the misguided ideological extremism of Mao Zedong.

One can imagine endless debates about which people infected others. But from a theoretical standpoint it is much simpler and cleaner to explore the possibility that the youthful unrest which sprang up the world over arose from a source or sources at once common to all manner of economic, historical, political, and cultural traditions, and particular to none. It is simpler and yet quite remarkable to think that such a broad array of societal differences is irrelevant to the timing of waves of youthful unrest, and concomitant waves of parental alarm.

THE REPRESSIVE EIGHTIES

Another political wave has swept the world in the 1980s. It is a conservative period. The position of traditional elites is consolidated. As Batra (1987) points out, wealth is being concentrated, both by restricting money supplies and by resisting redistribution of available wealth. While the rich get richer, the poor get prison in record numbers (to use Reiman's phrase; 1984). Xenophobia mounts.

The concentration of wealth is reflected in elderly circles of political leadership, represented for instance by Ronald Reagan in the U.S., Andropov/Chernenko in the Soviet Union, and participants in the 1930s Long March in China. It is not necessarily that an elderly elite conspires to hold onto power. For a while at least, electorates who are given options in Europe and North America repeatedly elect conservative governments. Opposition to conservatism seems to wane throughout class structures, so that, for example, the Labour Party in Britain and the Democrats in the U.S. are confronted by major losses of working-class support. Traditional means to upward mobility rise to prominence—for instance, the examination system for getting into Chinese universities, or so-called competency-based education in the U.S.

Melossi and Pavarini (1981) observe that over recent centuries, in Europe and in North America, such conservatism places a premium on individualism. People are said to rise or fall on their own initiative. The poor remain so because they are lazy or stupid, while wealth and power are attributed to effort and intelligence. From this perspective there is no point in economic redistribution. Rather, in the "natural" scheme of things, people will achieve as much as they ought through self-discipline. Cleaning up social messes is ultimately a matter of personal hygiene. If there is anything good to be done with the poor and homeless, it is a matter of purging impurities from and instilling discipline in their bodily functions. In this conservative ethos, our path to success lies in purging our bodies of drugs, in eating right, in keeping physically fit, and in working hard.

Nothing represents the combination of repression and xenophobia so directly as industrialized nations' wars on drugs. The major villains in this drama are darker-skinned people from the Southern world. Punishment focuses on purveyors of drugs of Asian and Caribbean descent in Europe, and of Central, South American, and Mediterranean descent in the U.S. Never mind that native white elites, as in the Drug Enforcement Administration and the Central Intelligence Agency in the U.S. (Chambliss 1987), are probably more central to drug trafficking than the outsiders can possibly be. The point, as made in the preceding chapter, is that Southern people of color are convenient targets for expression of the xenophobia and repression white Northerners feel on other grounds.

Beyond wars on drugs, the punishment of offenders has risen, not only in the U.S., but in China (see, e.g., Tifft 1985), and even in countries renowned for penal restraint, such as the Netherlands and Norway. This includes the dramatic resurgence of capital punishment in China in the mid-eighties, the resurrection of executions in the U.S., and the near reinstatement of the death penalty in such countries as Britain and Canada.

Like the youth unrest and parental fear of the sixties, the conservatism of the eighties occurs across a wide array of political, economic, social, and cultural traditions. I think that the remarkable coincidences of the sixties and eighties provide a key to a new way of understanding how societal levels of violence, including crime, wax and wane. Mind you, my own understanding at this stage is at best rough and preliminary. Nevertheless I will outline some of the patterns which begin to emerge as I try to place the coincidences of the sixties and eighties in a larger theoretical and historical context.

THE MECHANISM THAT DRIVES TWENTY-YEAR CYCLES OF SOCIAL UNREST

As far as I can see, the first of a pair of waves of unrest covers a period roughly twelve to twenty-five years after the end of a major war—that is, a period beginning with the onset of adolescence of postwar babies and ending when postwar babies displace prewar babies as parents (that is, shortly after the end of the postwar baby boom). The second wave, that of conservatism, swells roughly twelve to twenty-five years after the first wave has crested. The leadership in the second wave consists of persons old enough to have held some at least lower-echelon leadership position during the war years—military officer rank or privileged civilian position. That is, the elite status of the conservative leadership was established during the war.

The first wave

Among the countries suffering waves of youthful unrest and crime in the 1960s, all had a substantial proportion of their young men go to war in the 1940s. In China's case, the Second World War, which they shared with other countries, merged into the War of Liberation from the Nationalists, which ended in 1949. It is noteworthy that countries known for low incarceration rates and freedom from youthful unrest and delinquency in the 1960s—Costa Rica (Montero 1983) and Switzerland (Clinard 1978)—were neutral during World War II.

In those nations involved in the World War II, the war's end brought veterans home to father children. Women were in many places sent back to their homes to make way for male employment in peacetime economies, although the tide of women's participation in the paid workforce could only temporarily be reversed. Birthrates soared, while, in the Northern Hemisphere particularly, infant mortality fell.

While many postwar economies expanded dramatically, they did so by extending rather than converting from wartime modes of production. Mills (1956) was an early, prominent observer of this phenomenon. The greatest industrial giants, such as General Motors, had been built on major wartime military state subsidies. It was not only that military production and sales grew in the expanding cold war. Other leading industrial products and production methods were also direct outgrowths of World War II. Automobiles, for example, were built much like the tanks that previously had come off the assembly line—gas-guzzling monsters framed in heavy-gauge steel. Large-scale hierarchical production systems dominated the economy. As the president of General Motors, whom Eisenhower appointed Secretary of Defense, put it, "What's good for General Motors is good for the country." While levels at which American and other soldiers fought and died on battlefields dropped off, the leadership that emerged victorious from the war ruled with a military mindset, by military methods, whether the leadership was revolutionary as in China, or conservative as in the U.S.

Into these military zones came the baby-boomers. As Bateson (1979) put it, reproduction is a form of entropy. Much as parents like to think of children as extensions of themselves, reproduction serves to mix up gene pools and vary familial genotypes. What is true of genotypes applies even more to phenotypes in postwar baby booms. When the baby-boomers grow up, they are too many and too varied simply to succeed to the same status jobs as their parents. The problem is like that of patriarchal feudalism, where no more than one son can inherit as much land as his father holds unless other sons inherit less land from their fathers. Where wealth, power, and status matter—as they very much mattered in the postwar military states—wealth, power, and status matter relatively. If everyone gets richer, for instance, that signifies inflation and means to the wealth-conscious that everyone is in fact poorer. So when the baby-boomers grew up to assume their place in the economy, it was clear that the only way they could do as well as their parents was for the wealth, power, and status of their families to expand faster than the wealth, power, and status of other families. This pressure could be expected to be felt most from top to bottom in nations that had emerged victorious from the war. At any level at which families had participated in the war and felt the taste of winning, it would have been bitter indeed to see their children losing the ground they had fought to gain.

Adolescence, by definition, marks the crucial period when the successful launching of children into adulthood hangs in the balance. Throughout nations of war victors, the adolescence of postwar babies should have been a time of major anxiety for parents and children alike. Wherever the prospects of the children for being as rich, powerful, and respected as their parents hung in the balance, parents confronted possible failure at childrearing, while children confronted possible failure at maturation.

This tension flows out of people in varying mixes along three possible paths. Along one path, families or nations fight to take wealth, power, and status from others. Along a second, parents and children blame each other for impending failure. Along the third, people obtain relief by discovering that something matters more than wealth, power, and status.

All three of these paths were followed in the sixties. I see no basis for predicting which path will be taken by whom, except that the third path is least likely to be followed by the biggest winners and losers of war. Call it cognitive dissonance or inertia: families and nations which have gained the most wealth, status, or power from war will have an especially hard time giving it up. So will those to whom the threat of death from war, famine, or pestilence has become most imminent during war or in its wake. Those most likely to give up concern for wealth, power, and status should be those for whom pre- and postwar existence seems pretty constant. Neutral nationals have already been mentioned as falling into this category. So, I think, do members of relatively isolated, participatory, small-scale, self-reliant political economies. Among big winners of war, it occasionally happens that such political economies evolve over twenty-odd cycles of twenty-odd years. This appears to be the case among Norwegians, whose Viking Empire reigned supreme in Europe as late as the twelfth century, and who finally stopped sending armies abroad to fight after the Napoleonic Wars in 1820. Twenty years thereafter, one more wave of unrest and incarceration peaked. By the end of the nineteenth century, incarceration rates had been cut from 175 per 100,000 population to the current rate of roughly 50 per 100,000. Grassroots democratization of the political economy evolved noticeably during this transition (Pepinsky 1987b).

On a smaller scale, within nations, children of families whose middle-class standing or elite standing had remained largely unchanged from before the war would have been expected to be most likely to have dropped more enduringly from the game of acquiring or holding wealth, power, and status in the 1960s.

These are of course matters of probability rather than certainty. Anywhere at any time, it is both possible and highly improbable for those who have been involved in war in one generation to refrain from violence as they raise their own children.

The prospects of youthful rebellion against their own parents are greater if prosperity continues during the first wave, as it did much of the world during the 1960s. The first wave following World War I, in the 1930s, brought worldwide economic collapse, and rebellion was a luxury youth could ill afford. (Batra, 1987, argues that a depression might have been expected in the 1960s, too, and that since that depression was avoided, we are due for an economic collapse of unprecedented proportion in the 1990s.) On the other hand, youth in the thirties and forties were ripe for military mobilization; the Nazi Youth Movement is a prime illustration.

There is a danger of war recurring during the first wave of unrest. To politicians, war offers a way to discipline youth, to offer youth a measure of power and status, and potentially to expand the wealth available for domestic distribution. If the war is fought only on foreign soil, the war provides an outlet for parental violence which partially substitutes for imprisonment or other punishment of youth at home. That is why American incarceration rates have leveled off or dropped in this century only during major foreign wars (Pepinsky 1987b).

For now at least, I can see no clear explanation of why economic expansion rather than collapse occurred in the 1960s, nor why so many countries stayed out of war. China did enter a civil war known as the Cultural Revolution. The U.S. did invade Vietnam with a vengeance, although, from a domestic standpoint, a more limited number of young Americans were engaged in the Vietnam War at any one time than were mobilized in the military and in military production at any moment during World War II. So the precise form that the first waves of postwar unrest take remains undetermined, and this indeterminacy may amount to room for political choice to avoid certain forms of violence and privation.

War itself, then, is a culmination, but not necessarily *the* culmination, of waves of unrest following preceding wars within a generation. I would not expect the first wave to end and be followed by the second until a short while after the war is over. It happens that the wars I have looked at and thought about in recent history have been short-lived. I have yet to explore what happens when a Hundred Years War occurs. Perhaps so protracted a war actually consists of smaller overlapping wars involving different locales at various times. It does seem clear to me that a society caught up in its own cycle can have the cycle interrupted or perturbed by another society's cycle, as when one nation invades another, previously independent nation. It appears to me that involvement in World War I might have interrupted and modified the periodicity of American waves of unrest. Batra (1987) sees the U.S. Civil War as disrupting American economic cycles. If I am correct, the disruption goes the other way round—that by the mid-nineteenth century the U.S. had become independent of European economic cycles, and became reentangled during World War I. Such complexities are well worth exploring, but beyond the scope of this preliminary outline.

The second wave

As the children of the first wave "mature out" of war or other youthful unrest, a political symbiosis occurs between them and their parents. While the unrest might have provided a new political outlook for future reference, what is most apparent in the wake of the first wave is disarray. Prospects for employment are now uncertain for young and old alike. It is not really clear who owns what or who ranks where. Given the high probability that people who have believed in wealth, power, and status, and their children will continue to believe so, there now occurs a longing for order among both parents and children of the first wave of unrest.

The parents of the first wave of unrest continue to believe that they, better than their children, know how to create that order. Moreover, as the parents approach the end of their lives, they become more anxious to establish an enduring legacy, an earthly monument to the value of their lives.

The children, meanwhile, tend to grow weary of war and rebellion, and to long for a security that they themselves have been unable to create. Now, if their elders promise to manage their affairs and restore security, they are ready to be managed.

For both generations, there is inertia, for, having experienced change and disorder, they long for a restoration of some illusion of past order. If either generation had once felt blessed with special creative energy, now that charisma begs to be routinized (as in China in the late 1970s; see Pepinsky 1982a).

In the U.S., we know the two most recent of these second waves as the Reagan Era of the 1980s and the Eisenhower Era of the 1950s, when senior father figures have been elected President by overwhelming margins. (I use the term "father figure" advisedly. Even Prime Minister Thatcher plays the role of a patriarch—plays the stereotype of the male power figure rather than of the female nurturer.)

World War I left a similar period in its wake, noted for a concentration of wealth and conservatism comparable to the 1980s (Batra 1987), for repression of political deviation (see Zinn's descriptions, 1980), and for special punitiveness in criminal justice (see Kramer 1982 on parallels between the 1920s and the latter 1970s and the 1980s) comparable to the 1950s and to the post-Watergate era. These are times where meanness of spirit and narcissism are prominent. The security the generations want eludes them. The younger adults tend to live for today, while their elders tend to try to hang on in desperation.

Overlapping progressions

As long as wars do not last more than half a generation, nor new foreign entanglements perturb a previously established cycle, the second, conservative wave of one cycle is followed by the first wave, that of youthful unrest, in the next. A new generation of leadership replaces the elders of the second wave just as the first wave swells. The leading international personality heralding the new generation of leadership after the Eisenhower Era was John Kennedy; the leading international personality heralding the new generation of leadership after the Reagan Era is Mikhail Gorbachev. This is the time of greatest likelihood of political revolution.

The mix of a new generation of leadership and alarm by and toward youth is a particularly volatile mixture. It makes the political position of the leadership doubly tenuous. The tenuousness of their position makes demagoguery and warfare not inevitable, but especially appealing. Zinn (1980) gives vivid descriptions of the strength of this appeal as leaders drew the U.S. into the Indian Removal of the Jacksonian Era, the Mexican War, Secession and the Civil War, the Spanish-American War, World War II, and Vietnam.

The two waves have at least in this century in Europe and in North America lapped over and followed one upon the other with depressing regularity. What appears to some to be progress, as in the 1960s for the left and in the 1980s for the right, turns out within a generation to be ephemeral. Notably for criminologists like me, criminal punishment especially of underclass young men either rises steadily or is interrupted by wars in which underclass young men—together with underclass women and children who happen to live at the battlefront—are the first to die. There can be no enduring freedom from punishment and fear of crime or war, unless, as in Norway by the mid-nineteenth century, the cycle is broken.

CYCLES OF UNREST AND VIOLENCE AS CHAOS

This way of looking at the generation of war, punishment, and other violence is as radical a departure from conventional social science as relativity is from Newtonian physics. More precisely, this cyclical perspective is to social science as chaos theory is to established physics and mathematics (Gleick 1987).

Strange attractors

People are born, move through adolescence, and age continuously. Why, then, does intergenerational conflict occur in waves rather than constantly?

I have already related the cycles of unrest and violence to the occurrence of war. In a larger sense, the cycles can be explained in terms of what chaos theorists call "strange attractors" (Gleick 1987:119–53).

Chaos theorists use Poincaré maps to depict strange attractors. The illustrative Poincaré map in Diagram 1 is taken from Gleick. Diagram 1 traces the path the end of a pendulum takes when the pendulum is swung not just back and forth, but in a circle. Seen from the side, the path becomes a chaotic bundle of pendulum cycles, up and down, back and forth in no apparent pattern.

The Poincaré map, the kind some chaos theorists would computer-simulate, is a cross section of the pendulum's flow. Each dot on the map is a pass of the pendulum. Where each dot will appear in relation to any preceding pass of the pendulum is completely unpredictable. But after a number of passes, in the midst of this chaos, the dots as a group begin to form a pattern. And, almost magically, the pattern becomes sharper—gets filled in—as the flow of the pendulum continues. This pattern is an instance of strange attraction.

As this pattern is to the flow of the pendulum, so social and

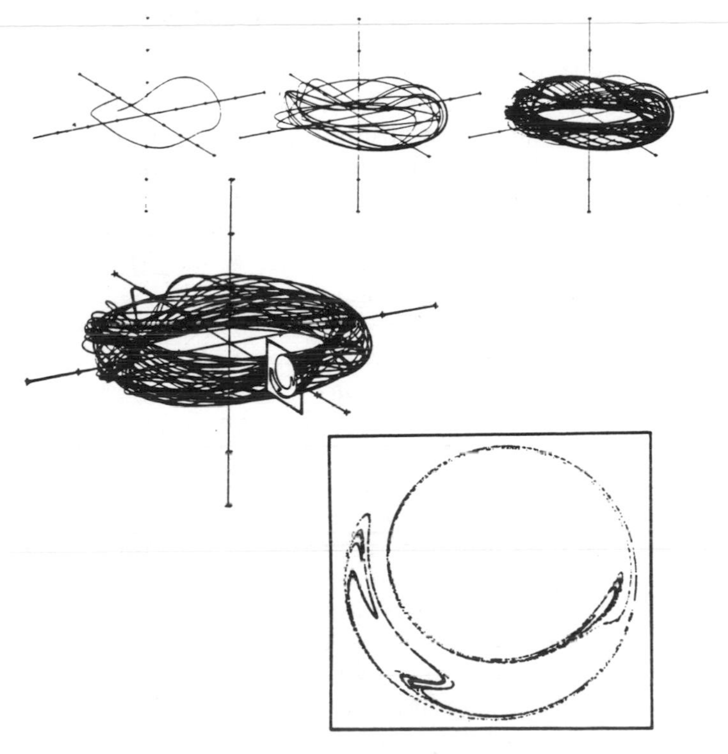

Diagram 1: Exposing an Attractor's Structure

The strange attractor above—first one orbit, then ten, then one hundred—depicts the chaotic behavior of a rotor, a pendulum swinging through a full circle, driven by an energetic kick at regular intervals. By the time 1,000 orbits have been drawn, the attractor has become an impenetrably tangled skein.

To see the structure within, a computer can take a slice through an attractor, a so-called Poincaré section. The technique reduces a three-dimensional picture to two dimensions. Each time the trajectory passes through a plane, it marks a point, and gradually a minutely detailed pattern emerges. This example has more than 8,000 points, each standing for a full orbit around the attractor. In effect, the system is "sampled" at regular intervals. One kind of information is lost; another is brought out in high relief.

humanly constructed physical structures are to the intergenerational flow of human life.

Consider first the chaos of human existence through time. As mentioned, reproduction is inherently entropic, both physically and socially (Bateson 1979). It is now believed that everyone alive today is descended from a single woman, probably an African, about 8,000 generations ago. The same genetic probabilities indicate that 90 of every 100 surnames which existed in Spain about the time Christopher Columbus reached Hispaniola would have disappeared today (Tierney et al. 1988). In 1657, Pierre Tremblay married Ozanne Achon in what is now eastern Canada. Today, they number 180,000 Tremblays among their North American descendants (Curran 1988). Against the background of as few as eight or nine generations of descent in the U.S. since its founding, combined with all the migration, even a limited examination of a half a century or so of vertical mobility like Mills's (1956), let alone a study of mobility between two generations, amounts to taking a virtual cross section of humanity.

In her workshops on imaging peace, Elise Boulding puts it this way: There are people living today more than 100 years old, and people just born who will live 100 years; hence, our living present-day existence is a 200-year window. White North Americans are especially unaccustomed to thinking in longer terms, but if, as many of the world's people do, we consider the dynamics of human evolution in terms of how succeeding generations of humanity modify human existence, chaos is the overwhelming reality of human inheritance.

Think of each pass of the pendulum in Diagram 1 as a line of descent in a human society. Where one line of descent emerges in relation to the next is a matter of chance. Given just a few generations, it is the legacy of any common ancestor to deposit people in perhaps widely separated, independent places. And yet, as people get deposited helter-skelter in a social system, physical and social group patterns emerge as strange attractors do.

If chaos theorists get excited when a strange attractor emerges in the swing of a pendulum, or in a simulation of Jupiter's great spot, small wonder that people become enchanted by the elaborate physical and social structures into which they find themselves born. Problems emerge, though, when people try to hang onto these structures or put concerted energy into creating new ones—from pyramids to monetary, class, and national systems—so imposing and rigid that they remain as monuments to confer an earthly immortality on their builders. A series of contemporary writers describes this conservatism as a source of violence: Brock-Utne (1985) and Marongiu and Newman (1987) call it domination; Foucault (1977) called a version of it discipline; and Shoham (1985) describes it as idolatry.

Trying to hang onto structure or to build monuments is like trying to push the pendulum in one direction each time it passes through the Poincaré map. Without any push, of course, the shape of strange attractors will gradually change as one draws successive Poincaré maps from along the pendulum's cycle. The irony is that if you push the pendulum, the strange attractors will change more abruptly. In fact, chaos theorists have found that the faster you accelerate a chaotic force, the more rapidly it changes, oscillating exponentially in a process chaos theorists call "the butterfly effect" and "period doubling" (Gleick 1987:11–80). Acceleration shortens the period. So it is with trying to force people to stay in or invest in one structure: the greater the force applied, the more turbulent the aftermath. World War II culminated in an ordering of the international political economy with unprecedented concerted force, and, in the aftermath, the turbulence of the sixties and conservative repression thereafter engulfed the world on an unprecedented global scale as well.

The Menger sponge

Encountering life in human strange attractors yields another reaction besides the urge to perpetuate structure. This is the reaction Ruddick (1980), the radical feminist, calls "maternal thinking." Buddhists call it "compassion." Jews call it "charity." Christians call it "love." It is the feeling that one most thrives in following one's own path by supporting one's neighbors in following theirs. This is what I have called "responsiveness" in chapter 2. Responsiveness lies along the third path for release of violent tension as described above in the section on "the first wave."

This drive to be responsive, and a corollary alarm at unilateral imposition or perpetuation of structure, must be human traits of overwhelming strength. Otherwise it would be impossible to imagine survival of the human species for more than a single generation. It is what Bateson (1979) describes as the negentropy which balances the entropy of reproduction. Compassion is the social counterpart of homeostasis in the individual. For human survival across generations, compassion has to go much further than women carrying children in their wombs for the first nine months of life. Thereafter, without having earned or achieved their way by dominating others, people need years of care, feeding, and education before they become capable of bearing their own children. Human birthrates are low and slow enough that one of every three or four children, if not more, must survive to reproductive adulthood for species survival. Cockroaches may breed and mature fast enough, and be hardy enough, to survive without compassion, but the human circumstance is radically different from that of the cockroach.

Compassion is like a resonance among the swings of the pendulum, which allows a structure among them to emerge and to evolve gradually rather than disappearing in turbulence. In terms used by chaos theorists, compassion is a catalyst for strange attraction in human activity. At the micro-level, the resonance manifests itself as enduring relations of mutuality and respect. At the macro-level, the micro-stability manifests itself as ties which cut across groups, classes, factions, families, and nations. Crosscutting ties, embodied in Shakespeare's myth of Romeo and Juliet's love, have been found by anthropologists to be a condition of societal peacefulness (Van Velzen and Van Wetering 1960).

Consider the form of a social structure where micro-relations become established and endure even as they cut across place and social identity. On the one hand, this structure has lines running so long in so many directions that no single group or habitat has any uniform substance. This is the way Jacobs (1961) describes safe, thriving neighborhoods in the heart of large American cities. There short, twisting streets predominate over major thoroughfares, so that people circulate variously and every which way. Old buildings mingle with new, small with large. Residence mingles with commerce. Income level of neighboring residents and business owners varies freely. So do ethnic identities. Social ties are so long and so manifold that no one group has enough mass to dominate or displace others. Everyone takes part in caring for the neighborhood because the task can be neither left to nor appropriated by anyone else in particular. By containing so much variety within, each block in the neighborhood is on average like any other, which the Brantinghams (1975) find to be the optimal condition for minimizing urban crime (as inferred in Pepinsky 1980:109).

Chaos theorists describe a physical solid with these characteristics as a Menger sponge (as shown and described in Diagram 2, taken from Gleick 1987:101). The Menger sponge has infinite length, but takes up no space and no volume.

The Menger sponge has more strength and resistance to environmental impact than any more solid form of the same substance. The strength and resistance of length in place of space and volume describe why my woolen gloves—with all those holes—keep my hands warmer than my thick and substantial leather gloves. So, too, the longer and less substantial a social structure, the more resistant the structure to war, famine, and pestilence. Cycles of war and unrest dissipate as the strength of compassion displaces the substance of building and maintaining idols. Just as one block, then one neighborhood looks more like another in the chaos of Jacob's (1961) healthy neighborhood, so the structure of one year, one decade, one score of years and beyond looks stabler the more freely social ties cut across one another in any cross section of humanity.

Scaling

The Menger sponge is an infinite regress of cubes surrounded by twenty smaller cubes, reflecting another principle chaos researchers commonly find in nature and in computer simulations, called "scaling." A pattern that appears at

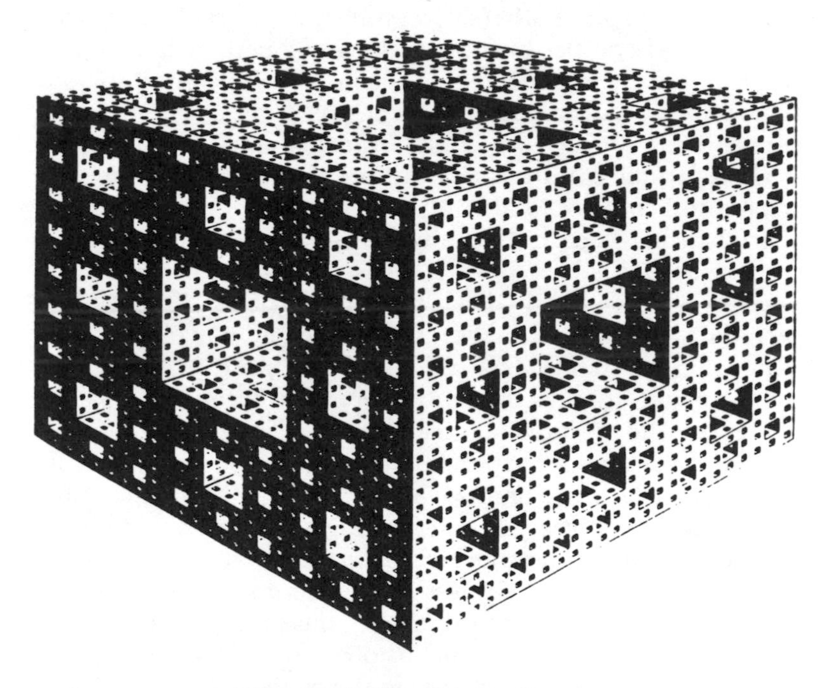

Diagram 2: The Menger Sponge

Constructing with holes. A few mathematicians in the early twentieth century conceived monstrous-seeming objects made by the technique of adding or removing infinitely many parts. One such shape is the Sierpiński carpet, constructed by cutting the center one-ninth of a square; then cutting out the centers of the eight smaller squares that remain; and so on. The three-dimensional analogue is the Menger sponge, a solid-looking lattice that has an infinite surface area, yet zero volume.

one scale recurs at a smaller scale, and, within that smaller pattern, recurs at a still smaller scale, and so forth.

> Scaling also became a part of a movement in physics that led . . . to the discipline known as chaos. Even in distant fields, scientists were beginning to think in terms of hierarchies of scales, where it became clear that a full theory would have to recognize patterns of development in genes, in individual organisms, in species, and in families of species, all at once. (Gleick 1987:116)

My initial explanation of the waves of unrest and repression in the 1960s and 1980s was cast in largely psychological terms—of how parents and children felt and related to one another. This crossing of system levels is typically frowned upon in today's social science, and yet it seems to me to be a fair description of social dynamics. "Just say no to drugs" both characterizes a pretext and language for, and implies a common approach to, for instance, addressing racist waves of incarceration in Europe and North America, waging white imperial warfare against Southern peoples of color who try to operate outside the established world economy, launching local police and school campaigns, arousing anxiety felt by parents over the adolescence of their own children, and experiencing personal turmoil and struggle many people feel about their own propensity to dependence and addiction. There is no more to be gained from trying to figure out which of these scales takes precedence over the others than there is in trying to figure cause and effect as between chickens and eggs. There is no point in distinguishing whether we anthropomorphize social structures and forces or are reductionist in explaining psychodynamics. What matters is that what happens at one level parallels what happens at others, and that if and when change is somehow managed at one level change at other levels will be entailed, whether as cause or as effect. A change at one system level cannot endure without corresponding changes at other levels (for a description of this phenomenon in the case of controlling crime in the U.S., see Pepinsky 1980:143–67).

Brock-Utne (1985) has written a particularly inspired and compelling radical feminist analysis of how scaling occurs in matters of violence and peacemaking. From issues of raising her own children, to her own role as a teacher, to larger studies of childrearing and education, to how science is done, to parental attitudes toward soldiering and militarism, to issues of military production, to battering women at home and raping women in war, to economic inequalities of gender and class, to international relations, Brock-Utne shows how each level mirrors the others. Take away the jargon and isolation of academic and professional specialization, and precious few models of good and bad, the practical and the impractical, recur at all social levels. It matters not which is the reality and which the metaphor. Things that are defined as real are real in their consequences, as W. I. Thomas put it, and the limits of our imagination in one social realm seem to limit our imagination in others.

On the positive side, to be capable of seeing one realm of life in a new way implies the capacity and the probability of seeing others in new ways as well. What enough of us can see and do for our personal selves has the potential of being seen and done among nations and huge organizations. Therein lies whatever hope we have for breaking the cycles of violence we now suffer.

TRANSLATING PRINCIPLES OF CHAOS INTO SOCIAL AND PERSONAL POLICY

Three major implications of this chaos theory of violence and unrest, for dissipating cycles of war, unrest, and repression, specify further what it takes for responsiveness to be organized at a societal level. Each of the implications is a condition necessary among other things to reducing crime and punishment (as occurred in Norway in the latter half of the nineteenth century).

Wherever several or more generations of people drop out of the war cycle: First, arranging how decisions are made takes precedence over deciding outcomes. Second, cultivation of one's own compassion takes precedence over changing anyone else. Third, for societies as entangled as the United States is in cycles of violence, the kind of peace Norway has achieved will at best take generations more to achieve.

"How" is more important than "what"

The essence of causing turbulence by building substantive structure is determining where people will be or what people will do before they confront issues and decide these issues for themselves. In Christi's (1977) terms, in matters of conflict this amounts to taking ownership of the conflict away from the disputants. In Wilkins's (1984) terms, predetermining substance amounts to a denial of democracy. In the terms of Tanzania's first president, the brilliant political theorist Julius Nyerere (1969), setting substance amounts to a denial of *ujamaa* socialism (literally, the socialism of familyhood, in Nyerere's view an East African tradition), and he writes of Tanzanian development:

> Any model which is drawn up should just be a guide which draws people's attention to the decisions which have to be made by them; each village community must be able to make its own decisions. (Nyerere 1969:271)

It has been Wilkins's (1984) genius to identify and describe the first condition—"how" is more important than "what"—in the context of criminal justice. Wilkins achieved international eminence for constructing a scale for predicting delinquency, then had the courage to reject the enterprise of prediction on moral and practical grounds. As a designer of parole guidelines—the forerunner of sentencing guidelines—Wilkins stressed that guidelines should not determine decisions. As large numbers of events fall into normal distributions, or in the manner pendulum swings reveal strange attractors, so Wilkins observed that large numbers of decisions fall into implicit patterns. Thus, although parole board members were inclined to believe that they decided each case individually, a statistical analysis of a board's decisions would inevitably reveal that the board members followed a policy in which at most four or five items of information explained about 85 percent of the variance in decisions whether to release inmates.

The guidelines Wilkins helped design were simply expressions of this implicit policy. They served not to prescribe what the board should do, but to describe what the board had done.

Thereafter, the board could be expected to continue conforming to its policy most of the time. But, more important, the guidelines made it possible for inmates and board members alike to challenge the wisdom of doing business normally with respect to particular cases. As a matter of principle, Wilkins called on board members to depart from the guidelines in 15 percent of their decisions. Each time the board departed from the guidelines, they were required to give publicly available written reasons for doing so. Inmates were encouraged to challenge both the guidelines and the departures at any time. The board itself was required to review its exceptions annually, with an eye to finding reason to reformulate the guidelines themselves.

Wilkins saw the guidelines as a device for making decision-makers accountable first and foremost to their subjects and to themselves, and for forcing decision-makers constantly to revise the substance of their decisions in light of feedback from their subjects. What Wilkins (1984) calls "consumerist criminology" is an extension of this idea of public accountability and accommodation to criminal justice decision-making overall.

The Wilkins model turns conventional regulation on its head. Rather than regulating by making decision-makers subject to superiors' enforcement of restriction on their discretion, Wilkins would enhance discretion and make discretion accountable to and susceptible to the influence of those affected by decisions (see Pepinsky 1984 for an application of these principles to policing).

Tocqueville (1945 [1840]) foresaw the possibility that even in the formally democratic U.S., people who were preoccupied with material comfort might be all too prepared to let experts and officeholders make their decisions for them, a form of self-imposed "depotism." A belief in subordinating oneself to substantive structure amounts to a belief in domination (Brock-Utne 1985). Even power elites in such a belief system see themselves not so much as giving orders, as passing along orders they were somehow bound to give. Lower-class subordinates in such a system believe that they are not merely subject to orders, but that they are entitled to give orders to subordinates of their own—the husband to his wife, the wife to her older children, bigger boys to smaller girls, while the smallest kid in the system gets to kick the cat or to make a mess of the physical environment, or perhaps waits quietly to grow old enough and big enough to have another, smaller person to order instead: the lightning rod effect described in chapter 2.

The obstacle Wilkins ran into is that those accustomed to living by domination can make a mockery of any formally democratic system imposed upon them. Parole boards stopped getting around to reviewing their exceptions and revising their policies accordingly. By now in successor guidelines, as in the new federal sentencing guidelines, the guidelines are simply set by administrative fiat; even the decision-makers get left out, to say nothing of their subjects. Henry (1984) has found that organizational form is no predictor of disciplinary practice. A formally participatory cooperative can inflict hierarchial sanctions on members, while a formally hierarchical organization may regulate employee behavior nonhierarchically.

From the standpoint of chaos theory this perversion of substance is no surprise. The structure of the strange attractor, remember, is independent of the flow of people through it. If the flow of action runs independently of established structure, still less will the flow of action conform to newly imposed structure. All that can be said is that the grander the scale and the greater the force with which structure is imposed, the greater and more turbulent the waves of violence which will lie in its wake—waves whose period is constrained by the rise and fall of generations rather than by structural form. It may be that the dissipation of violence requires that those most directly affected by decisions make the decisions as contingencies arise. But knowing where to go does not in itself tell us how to get there.

Self-cultivation precedes imposition on others

> Viable socialist communities can be established only with willing members; the task of leadership and of Government is not to

> try and force this kind of development, but to explain, encourage, and participate. For a farmer may well be suspicious of the Government official or party leader who comes to him and says: "Do this"; he will be more likely to listen to the one who says: "This is a good thing to do for the following reasons, and I am myself participating with my friends in doing it." (Nyerere 1969:263)

> Therefore, the National Executive Committee [of the Tanganyika African National Union (TANU) Party chaired by Nyerere] . . . resolves:—
>
> 1. Every TANU and Government leader must either be a Peasant or a Worker, and should in no way be associated with the practices of Capitalism or Feudalism.
>
> 2. No TANU or Government leader should hold shares in any Company.
>
> 3. No TANU or Government leader should hold Directorships in any privately-owned enterprises.
>
> 4. No TANU or Government leader should receive two or more salaries.
>
> 5. No TANU or Government leader should own houses which he rents to others. (Tanganyika African National Union 1967: sec. 5A)

Of course, Nyerere has no more succeeded in getting TANU members and Tanzanian officials than peasants to live by socialism. On the other hand, Nyerere is credited with influence on account of his reputation for living by his own principles himself.

It is the message of students of violence like Brock-Utne (1985) and Quinney (1987) that whatever power we have to change others is limited by our power to change ourselves. A chaos theory of violence supports this premise, for three reasons.

First, the fundamental antithesis separating violent turbulence from peaceful stability is domination/subordination versus give-and-take, in a system where what goes around comes around. Success in subordination can only teach the virtue of domination, and making someone else do one's bidding or follow one's plan is—there's no way around it—subordination. The only option is to show that you enjoy living by compassion, and turn people on to sharing your own experience.

Second, even if some people deserve domination and their subordination makes us safer or happier, the domination inevitably extends to innocent bystanders, causing undeserved suffering and perpetuating violence. This is Smykla's (1987) powerful message regarding innocent family members of death-row inmates. Some Israelis are realizing that suppression of Palestinian terrorists is producing a hatred in Palestinian children that may threaten Israelis for generations to come (Grossman 1988). Newman and Lynch (1987) are led by endless cycles of vengeance (to them the root of all violence even at international levels) to conclude that a personal choice not to retaliate is the only way to break the cycle. Shylock's inability to take one pound of flesh and no more, in Shakespeare's *Merchant of Venice,* stands for the proposition that subordination cannot be limited to the giving of subjects' just deserts (Pepinsky and Jesilow 1985:120–30). This is the extension of the principle that those who are subordinated pass domination along where they can.

Third, since vertical mobility and migration are virtually limitless in the generations it takes to break cycles of violence, the people you treat one way because they belong to one class today may—by their descendants—end up in any other class later on. Does a proletarian or an elite member of society deserve preferential treatment today? The expectation of preferential treatment (or resentment of invidious treatment) that is instilled in the parent today may well crop up in the descendants at opposite ends of the class spectrum. This was the problem Chinese revolutionaries confronted in their own children, who as Red Guards mimicked their parents by revolting against parents and teachers during the Cultural Revolution (Pepinsky 1982a). In microcosm (as a matter of scaling), we see similar transitions in the maturation of individuals. What, for instance, of male children who learn to be perfectly obedient to teachers in school and fathers at home? Are they not implicitly taught to demand absolute obedience as adults from children, from wives, and from other subordinates (see Brock-Utne 1985)? What of the compliant prison inmate who upon hitting the streets goes berserk since there are no bars and guards to impose restraint? The corollary of learning that childhood or confinement entails discipline is learning that adulthood or liberty entails despotism.

We have the unfortunate habit of measuring compliance by immediate feedback. The quick and able subject learns to delay reprisal, to save it for times and places which will not be monitored, and for people who cannot manage to retaliate. At societal levels, as we have seen, waves of response to violence take decades to develop. It is not what happens in the structure here and now, but the impact of what happens here and now in structures yet to come that matters most in the generation of violence. The safest prediction is that what happens in one structure here and now could end up coming out virtually anywhere else in a social system. This is a secular counterpart of the Hindu notion of karma.

Personal cultivation means attending most to one's relations with one's nearest and dearest. It means, as Brock-Utne (1985) argues, that one does not know how to make peace among nations who does not treat one's own children or spouse or colleagues with respect and dignity. How does a teacher understand democracy who does not understand how to share power with students? How much does an expert on due process understand who gives orders to employees, students, or children without hearing or explanation? What message does an "authority" convey who preaches one standard and behaves otherwise? Think of how much more credible and powerful university departments would become if they applied their knowledge first and foremost to governing themselves: if historians dedicated themselves to avoiding repeating the mistakes they found in the past; if political scientists dedicated themselves to constituting model government with their staff and students; if sociologists were committed to constituting themselves as an ideal society; if criminal justice departments like mine were dedicated to perfecting justice among themselves; if professors of education were dedicated to perfecting their own teaching. . . . Imagine police forces where roll calls and training sessions were devoted to perfecting the law and order of the officers' own lives. Imagine a White House whose first commitment was to perfecting executive branch, then federal, then American democracy, rather than imposing "democracy" on Nicaraguans.

The chaos theory of violence implies that one's greatest impact on history will be with those to whom one most directly relates. The emotional power of face-to-face contact lasts more and is less garbled than operating at grander levels. The impact one has on a small child or even a derelict close at hand stands over the course of generations it takes to make peace to be inherited by persons as influential and powerful as the descendants of today's movers and shakers. Perhaps by moving in upper echelons and in abstractions one is capable of such wanton destruction and violence as to have a greater impact than by beating one's own wife, but when it comes to transmitting the power of compassion, the smallest moment at hand with one person has as much potential impact as the most dramatic, global activity.

Peacemaking takes a long, long time

A commentator recently observed that gauging economic progress by monthly trade deficit figures is like getting on a scale every five minutes to see whether a diet is working. In fact, a strategy that is working may in the normal course of events initially appear to fail. Chaos theorist William M. Schaffer has found, for instance, that after a successful inoculation campaign the rate of measles first spurts upward (Gleick 1987:315–16). So, too, those who respond to violence with compassion may—as in the story of the life of Christ—appear to open themselves to further victimization. It is certainly presumptuous to expect those who are getting hurt most by violence to make our peace for us—for battered women to stay quietly at home, or for targets of death squads not to resist, for example. But for those of us who are privileged enough to be able to afford compassion in our everyday lives with our families, neighbors, colleagues, and other associates, the question remains how much progress we may expect to witness.

One sad implication of the chaos theory of violence is that

the fruit of our own peacemaking efforts lies beyond our power of empirical verification. In rare circumstances where the history of a people who finally have made peace is documented, we may see that progress occurs, as in Norway, over some twenty generations. Indeed, whatever support for or refutation of this theory of violence exists for us to find lies in traces our forebears have left for us to see, not in the evidence we see of outcomes of our own actions.

This lays a theorist like me or like Quinney (1987) open to the charge of spouting religion rather than doing science. So be it. My reading of history teaches me that immediate empirical and supposedly practical results are ephemeral when it comes to controlling violence. The faith that makes me aspire to live by compassion is not blind in its origin, but is blind in the sense of succeeding—in theory—only if it is impervious to immediate setback or reward.

As a U.S. citizen, I can see that my country has probably reached its peak of global domination. I can foresee that for someone with my liberal inclinations, the prevailing political ethos of the coming decade will be a relief from the repression of the 1980s. I can foresee a Soviet-U.S. political accommodation and eventual Northern white military global alliance against Southern and Eastern peoples of color. I can see that the survival of our species is perilous, and could suddenly end at almost any time. But while human life and hope remain, my most favorable prognosis for the future of U.S. society is that it will become as peaceful in a few centuries as Norway became in the ninteenth century, and that global peace is more elusive still.

Perhaps the unprecedented is possible. Perhaps Americans and others now caught in cycles of violence can recognize the pattern and respond dramatically. Perhaps in a single generation socially aware parents will prove capable of treating their children democratically, and will give up on hierarchical structure and the building of monuments sufficiently to allow children to find a dignified, stable place in the adult world. If so, reason has a lot more power than I have yet seen. But even if the biggest impact of trying to understand, describe, and accommodate to cycles of violence is to help humanity inch forward on the millennia of roads to peace lying before us, the study of violence seems as worthwhile and rewarding as any other enterprise I can imagine.

CONCLUSION

For twenty years I have concentrated my professional life on trying to understand how to control crime and violence. I have seen criminological wisdom accumulate, become applied, and crime and violence mount as a result. I have repeatedly come close to deciding that nothing works, then found some experiment that seems to work, only to look closer and longer and become disenchanted. It has been even more disturbing, later in life, to begin reading more of the history of crime and violence. There I have seen ample precedent for the "new" initiatives that come along. This applies even to inspired political revolution.

Does nothing ever change? Of course it does. Recent reading about chaos theory (Gleick 1987) and economic cycles (Batra 1987) has helped me discern cyclical patterns of violence, and what I have learned in and about Norway has given me renewed hope that established historical patterns can be broken.

If this is how the progress of violence and responsiveness look at the societal end of the social spectrum, the question remains how the progress of violence and responsiveness are personally experienced. As a criminologist I am particularly concerned with how criminal "justice" is felt by its subjects. One criminal case has offered me a rare opportunity to understand a defendant's experience of violence and responsiveness.

4

A Criminal Defendant's Sense of Justice[8]

I have served with Bloomington attorney Guy Loftman as co-counsel in *State of Indiana v. William T. Breeden,* cause no. CR-86-75 (Daviess County, 1986–87). This case has shed new light for me on how a criminal defendant's sense of justice is satisfied or offended, not only because of my direct involvement in procedural complexity but because I happen to be close friends with the defendant and his family. I thus have data on the defendant's perspective that transcend what I would know from normal attorney-client contact.

As I reflect on the case, I can perceive the legal and social scientific ideology which confounds justice with speed and substance of case disposition. Case processing time and disposition ignore the defendant's reality that justice is at issue at many levels at once, not only in the court but in the defendant's relations in the outside community. At each of these levels, a sense of justice arises not from any particular outcome, but from the perception that the defendant or the advocate has influenced, without determining, others' intent. This influence would appear in empirical analyses only as an interaction effect rather than as the independent effect of any single party's input into an additive model. In standard additive models for analyzing court processes or dispositions, therefore, including models underlying the construction of sentencing or parole guidelines, greater justice would entail more unexplained variance. This is consistent with Aubert's (1959) thesis that verdicts must be seen as matters of chance—must be unpredictable—in order to be perceived as just (see also Pepinsky 1982b). But while apparent chance outcomes in additive analyses of court processes and dispositions are necessary for the processes and dispositions to be seen as just, my experience in *State v. Breeden* indicates that chance alone is not enough. In this essay I infer what elements of order are required for justice to be perceived, in a realm that legal and social scientific ideologies treat as an empirical void. Here it is argued that defendants will perceive the disposition of their cases to be just only when these previously invisible elements are taken into account and incorporated into the legal process.

This analysis stops short of prescribing remedies for injustices suffered by criminal defendants. The analysis merely suggests that social scientists and policymakers have to look beyond the data which appear in court records, indeed beyond what occurs in court at all, to appraise the significance of the trial process to defendants. If the larger purview of a defendant's sense of justice or injustice is not taken into account, the most meticulously designed systems for rendering justice to defendants may inadvertently—even unnoticed—go awry. It is speculated here that Bill Breeden's purview of justice represents the purview of criminal defendants generally. This hypothesis remains to be tested in further case studies.

This chapter begins with a description of the circumstances which gave rise to *State v. Breeden*. It then outlines the legal defense, and the proceedings through trial and sentencing. The ensuing analysis of the case begins with how issues of justice might be cast in a standard legal/social scientific evaluation, then turns to other issues of concern to the defendant. The study concludes with consideration of what justice means to Bill Breeden and other criminal defendants.

THE SIGN

Thanksgiving night 1986. Odon, the Southwest Indiana town of 1,300 where John Poindexter grew up and went through high school. John's father and uncle Poindexter had been vice presidents of the Odon Bank. The Poindexters have lived in Odon for nine generations. John's cousin Dicky Ray Poindexter worked for years on the staff of Senator Homer Capehart, who brought the mammoth Crane Naval Weapons plant and arsenal to the town's doorstep. John's father beat out John's uncle for the bank presidency. The two bankers tried to put each other out of business, but the bank thrived and the funeral home John's uncle established gained a virtual monopoly in the region. Dicky Ray Poindexter claims that the bank "is probably one of the largest Navy banks in the country. Naval officers from all over the country bank at the First National Bank in Odon, Indiana" (March 25, 1987 Deposition, p. 26). Senator Capehart appointed Odon native John Poindexter to the Naval Academy, and John achieved a more powerful position than any active naval officer in American history. But John had been disgraced, and, just two days before Thanksgiving, had resigned as National Security Adviser and been demoted.

There's a group that controls town politics. Membership circulates primarily in one church and in a pair of local clubs. Dicky Ray Poindexter is well connected in state Republican politics and is a member. He is proud that he has been an honorary deputy sheriff in Daviess County for thirty-eight years—the current sheriff being an old friend from Odon. John Myers is editor of the *Odon Journal,* president of the state Republican press association, and owner of the local license plate branch (a patronage franchise), which recently had $7,500 disappear from state accounts. John and the group like to play pranks. There is evidence that John paid boys 25 cents apiece to fill the back of his pickup with Democrat Frank McCloskey for Congress signs just before November's election. Joe Haskins owns a feed mill and sundry investments making him perhaps the wealthiest Odon resident. Kenny Hudson used to be high school president, is now principal of the elementary school at John and Walnut Streets. Kenny is a Republican member of the three-person Odon Town Board.

About six months earlier, Kenny Hudson had gotten the Town Board to agree to prepare to rename three-block long John Street "John Poindexter Street" by ordering four new street signs. The thought was that John could come home for Old Settler's Day in August for a street dedication ceremony.

But John couldn't make it, so the signs were left in the town workshop in hopes that John would come a year later.

The day after John Poindexter resigned, the national press had come to Odon asking whether the town was reconsidering the name change on John Street. John Myers was upset, and so were Joe Haskins and Kenny Hudson. Myers and Haskins got Hudson's blessing, took a John Poindexter sign out of storage, and wired it with no. 9 baling wire over the John sign at the corner of Walnut, while a TV crew filmed the event. Town Board President Frank Armstrong originally thought the sign should not be erected until the Town Board took official action, but eventually he went along.

Bill Breeden had also grown up in Odon. He was younger than John Poindexter; Bill's sister Ginger—a nurse in Odon—had been a high school classmate of the Admiral's. Bill had left eventually to get his master's in divinity and become a parish minister in Missouri. He had been a teenage evangelist and fundamentalist, but after meeting Phil Berrigan in 1979 Bill's life had changed radically. He and his twin brother and their families had become peace activists and were known as the Breeden Singers. For a couple of years they had lived near the Crane Naval Weapons Depot. It had become an annual event for Bill to camp out at the Crane entrance, fasting and passing out peace literature to Crane employees. Dicky Ray Poindexter had come to dislike this purveyor of "Communism" intensely.

The summer of 1986 Bill's wife, Glenda, had been a Witness for Peace in Nicaragua. She had returned in some despair. She and Bill had shared tears over the violence and destruction wrought by CIA/NSC-sponsored Contra terrorism in Nicaragua. Bill and Glenda were actively involved in raising material aid for Nicaraguan victims of Contra violence, to be gathered together and trucked to the East Coast for shipment the second week of December.

The Breedens had driven over to pay a Thanksgiving visit to Bill's mother in the nursing home right down at the foot of John Street. Starting home that night, they saw the John Poindexter Street sign. When they went to get a camera to take a picture of the sign, they heard about the news broadcast showing John Myers hanging the sign. They felt they could not ignore the sign or the political message of support for Contra terrorism it carried. For an hour or two they debated what to do.

THE CRIME

Bill took some wire cutters and cut the sign down. A note on the back of a piece of cracker carton was left taped to the sign pole, proclaiming that not all graduates of Odon High School honored John Poindexter. The note was signed "Midwest Liberation Front."

That same night, Deputy Town Marshal Doyle Webster's daughter told him she had seen the big beautiful sign. Doyle Webster went to see the sign and found it was gone. He was not concerned. He went to every Town Board meeting and knew that no one had officially used the sign. He figured that the sign had just been returned to town storage where it belonged. Two days later Joe Haskins told John Myers that the sign was gone, but neither of them cared to do anything about it. Kenny Hudson went right by the corner on his way to work, but took no notice of the missing sign.

The Thursday after Thanksgiving Bill Breeden went to a meeting on the Indiana University campus in Bloomington to publicize the material aid campaign. He had the sign with him. He had not been able to figure out what to do with it. He mentioned the sign to a local reporter. Bill's picture was taken with the sign. He parodied John Poindexter, taking the Fifth Amendment on whether he had taken the sign from Odon. He proclaimed that the sign was being held for $30 million ransom for the Nicaraguan people.

Dicky Ray Poindexter returned from a trip the day after the story appeared. He was furious, and called the prosecutor and the sheriff demanding that Bill be arrested and charged.

The following Wednesday the prosecutor got the judge to issue a warrant for Bill's arrest on a charge of felony theft "by taking the John Poindexter Street sign with intent to deprive the Town of Odon of the use and value thereof." Bill turned himself in the next day. A duplicate sign, made up as a gag, was left for Indianapolis Police to pick up from a phone booth at Eugene Street and Martin Luther King, Jr. Boulevard that Friday, just after Bill had left the city with a truckful of material aid for Nicaraguans. The real sign was left for the Daviess County Sheriff in Washington, the Daviess County Seat, the next weekend.

At Indiana law, "A person who knowingly or intentionally exerts unauthorized control over the property of another person, with intent to deprive the other person of any part of its value or use, commits theft, a Class D felony" [Indiana Code 35-43-4-2(a)].

A lesser included offense: "A person who knowingly or intentionally exerts unauthorized control over property of another person commits criminal conversion, a Class A misdemeanor" [Indiana Code 35-43-4-3].

The maximum sentence for a Class D felony is four years' imprisonment and a $10,000 fine, with a presumptive sentence of two years' imprisonment which can be mitigated or aggravated as the trial judge finds [IC 35-50-2-7(a)], while for a Class A misdemeanor the maximum penalty is one year of imprisonment and a $5,000 fine, with no presumptive sentence [IC 35-50-3-2].

THE DEFENSES

We moved, unsuccessfully, to dismiss the prosecution on grounds that Bill Breeden would not be prosecuted for an alleged theft but for the offense caused by his political statements thereafter. The local U.S. Court of Appeals has ruled that if the defendant makes a prima facie case of such selection or discrimination, the state has a heavy burden to prove that the prosecution would have occurred despite the speech. Otherwise, the prosecution must be dismissed as an infringement on First Amendment rights which amounts to an impermissible denial of equal protection of the law [*U.S. v. Falk*, F. 2d 616 (7th Cir., 1972)].

We had wanted to present evidence to the jury on this issue, figuring that they could better afford the political heat of dismissing a prosecution than the judge. The judge seemed to like this idea, until he found *Love v. State,* 468 N.E. 2d 519 (Indiana Supreme Court, 1984), cert. den. 471 U.S. 1104 (1985), which ruled that this issue was solely a pretrial matter for a judge to consider.

As the prosecutor argued to the judge, this case law arose out of the furor over Vietnam, which, the prosecutor argued, made *U.S. v. Falk* anachronistic. The judge denied our motion after a September hearing, finding, "The evidence discloses that the defendant was prosecuted for his actions in allegedly taking the sign, not his personal or political views." He then granted a motion *in limine* forbidding us to mention selective prosecution to the jury (although under the Indiana Constitution, art. 1, sec. 19, an Indiana criminal jury is a trier of law as well as a trier of fact).

At trial, we presented evidence that the Town Board had not officially decided to hang any John Poindexter Street sign, until after the sign Bill took was returned. We argued that Bill reasonably concluded that the sign was available for political use by citizens until the town decided to use it themselves. Private citizens had apparent authority to use the sign for political purposes as long as the sign was unhurt and returned when the town needed it. Meanwhile, as Bill reasonably inferred, taking the sign deprived the Town of Odon of no part of the sign's use or value. In fact, even if the town had officially meant to honor John Poindexter by letting Myers and Haskins hang the sign, Bill's actions helped publicize the town's message, thereby enhancing the use and value of the sign.

Finally, Bill and Glenda especially presented extended evidence of their beliefs that Nicaraguan lives were in imminent danger, and that after their discussion and sober deliberation Bill had decided that the sign must come down to scream "no" to carrying on the killing. Indiana Code 35-41-3-8 reads:

> (a) It is a defense that the person who engaged in the prohibited conduct was compelled to do so by threat or imminent serious bodily injury to himself or another person. With respect to offenses other than felonies, it is a defense that the person who engaged in the prohibited conduct was compelled to do so by force or threat of force. Compulsion under this section exists

only if the force, threat, or circumstances are such as would render a person of reasonable firmness incapable of resisting the pressure.

(b) This section does not apply to a person who: (1) recklessly, knowingly, or intentionally placed himself in a situation in which it was foreseeable that he would be subjected to duress; or (2) committed an offense against the person as defined in IC 35-42.

THE VERDICT

The trial began at 1:00 on Tuesday, October 20, 1987. The case went to a six-person jury that Friday just before 3:00 in the afternoon. Guy Loftman observed that unless the jury immediately voted acquittal, the timing was particularly hazardous for a defendant. In his experience, around 5:00 on a Friday afternoon jurors start to feel an unconscious urge to conclude their business and go home. Compromise verdicts are especially likely at this time. Jurors who believed in acquittal would give up.

At 5:30 the jury returned with a verdict of not guilty of felony theft, but guilty of misdemeanor conversion. We later learned that the initial vote of the jury had been five for conviction of the felony, and one for acquittal. We have no further information on jury deliberations.

THE SENTENCE

On grounds that the offense itself was minor but that the defendant was likely to recidivate, the probation officer recommended a sentence of one year in jail, with all but five days in jail suspended in favor of one year's probation and one hundred hours of community service.

The sentencing hearing was November 23, 1987. There Guy Loftman and I argued that since Bill Breeden had been acquitted of theft, the jury had found him innocent of depriving the Town of Odon of any use or value of its sign, and since the judge had ruled that Bill was not being tried for offense caused by his speech, and since no defendant in Daviess County convicted of a misdemeanor property crime the past year and a half had received a jail sentence greater than time served, that Bill should receive only a nominal sentence—a $1.00 fine. The prosecutor asked that Bill receive a jail sentence of fifteen days plus the one hundred hours of community service recommended in the presentence report. This was a major reduction from his pretrial bargaining position that if Bill pleaded guilty to the felony with a presumptive sentence of two years' imprisonment he would stand mute on sentencing. The judge read from a statement prepared before the sentencing hearing, declaring that Bill was not being punished for his politics, that no one "in authority" had influenced his decision, and sentenced Bill to a year in jail with all but eight days suspended, eighty hours of community service, and a year's probation. The Judge allowed good time credit of one day for each day served, and credited Bill with one day of the remaining sentence for his arrest, leaving three days in jail to be served beginning the Sunday after Thanksgiving.

I presented twenty-three signatures of people willing to serve parts of Bill's sentence. The judge said this was impossible, but that we could accompany Bill during his community service. Guy Loftman reminded the judge of Bill's early prophecy that he would get more time for taking the sign than John Poindexter would for his crimes, and asked that imposition of sentence be stayed until we knew whether John Poindexter was pardoned or convicted without receiving incarceration, providing a basis for reduction of Bill's sentence. The judge denied this motion.

Bill had never wanted to appeal this case under any circumstances. No appeal was filed.

A STANDARD ASSESSMENT OF JUSTICE

The framework for assessing whether juries and courts render just verdicts and sentences has not changed since I reviewed the literature over a decade ago (Pepinsky 1976a:44–56). Holding offense and prior record constant, "justice" is a null hypothesis which can be rejected by showing that the class to which the defendant belongs (e.g., black or white) is treated differently from that of other defendants, or that jurors or judges with some backgrounds convict or sentence differently from others.

By these criteria, it appears that the jury's verdict was just. Jurors were diverse by Daviess County standards; whatever potential biases might have cut against the defendant were offset by biases in the defendant's favor. In other words, we have no basis for supposing that a different jury would have found Bill either guilty of the felony or entirely not guilty. We were in complete suspense during the jury's deliberations. Some of the defendant's friends thought that Bill was likely to be acquitted; others thought that once he had admitted taking the sign, a guilty verdict was inevitable.

Bill Breeden, like other politically aware defendants I have known, placed great faith in having a jury hear his case. Jurors are certainly more insulated from political pressure than judges. Jurors can be open to hearing the defendant's side of the case. Jurors can do the right thing—which in Bill's view was to acquit him. The verdict came as a shock, but still Bill did not question the jurors' integrity. Rather, he questioned legal restrictions imposed on the jurors' deliberations—a matter outside the scope of normal inquiry into jury discrimination. Bill's belief in the jury's integrity stemmed from not being able to explain how jurors' personal predilections affected their deliberations. That is, as Aubert (1959) and I (Pepinsky 1976a:44–56) have argued, this defendant's perception of juror integrity rested on the element of unpredictability that had underlain determination of the verdict. Bill's residual perception of injustice rested on his retrospective inference that a conviction had been inevitable. But this inference was based on factors extrinsic to the characteristics of the defendant or of the jurors, and therefore lies outside the scope of standard inquiry into the justice of verdicts. I return to these factors below.

By the court's own terms, the sentence was unconstitutional, under law applicable in Indiana and the rest of the Seventh Circuit. The court ruled that Bill was not being prosecuted for his speech. And yet the Daviess County records showed that no one convicted of a property misdemeanor in the past eighteen months had been sentenced to more than time served or, in a few cases, had been given a concurrent sentence. Moreover, the jury had found Bill not guilty of intending to deprive Odon of any use or value of the sign. Bill received an extraordinarily stiff sentence. The court recognized as an aggravating factor that Bill had twice before been jailed—once for praying in the D.C. Capitol Rotunda, and once for contempt for failing to tell a Kentucky judge where to find his home-schooled children—indicating, together with Bill's own testimony, that he could be expected to be unrepentantly civilly disobedient if a future "gift" like the street sign presented itself. Bill's record indicated that where he found a chance to stage political theater without hurting a victim he would commit a technical violation of law. He would cause political offense. The judge acknowledged that Bill's crime had caused no "serious harm to persons or property." But, as we pointed out, Bill's crime caused no harm whatsoever except to egos of some members of a political elite. Thus Bill was constrained by probation from causing political irritation even by purely technical violations of law. He is punished not for what happened to the street sign, but for offense caused by what he said in the process. That offense led to an unusually severe sentence by Daviess County standards. In this traditional legal sense, the sentence was unjust.

Bill nonetheless feels great personal respect for the judge, and feels vindicated by the sentence. For him there is justice in the sentence. Once again, in order to explain this sense of justice, one must look outside the normal purview of legal inquiry.

A BIGGER PICTURE OF JUSTICE ISSUES

With family and friends living in and around Odon, Bill was well aware of larger community forces at work which circumscribed the court's discretion. From newspaper accounts and by word of mouth, Bill and many of the rest of us were convinced that members of the Republican elite in Odon—Town Marshal Dick Slaven, Dicky Ray Poindexter, and John

Myers to name three—were pushing and probably directly exhorting the judge to put Bill in jail for six full months (as by sentencing him to a year with an allowance for good time). Friends of Bill were apparently ready to engage in civil disobedience themselves if this sentence were imposed. In the words of one, "Bill wouldn't have gone to jail alone." Within this framework, the judge demonstrated considerable political courage on Bill's behalf. We cannot help suspecting that he clued in the prosecutor sufficiently to moderate even the prosecutor's recommendation. After imposing sentence, and over the prosecutor's objection, the judge allowed Bill to remain at liberty over the Thanksgiving holiday.

In the statement of reasons for sentence, the judge acknowledged input he had received—forty-odd letters on Bill's behalf—but insisted that he had received no influence from anyone "in a position of authority." This Bill and his Odon friends took to be a statement to Dicky Ray Poindexter and John Myers especially—seated as they were in the courtroom—that they could not bully the court into upholding their personal honor. The judge in fact expressed considerable sympathy for Bill's ministry, albeit noting that those who engage in civil disobedience must be prepared for punishment.

More than a hundred people had come to participate in a peace circle in front of the courthouse just before sentencing. Sixty or so of this group went with Bill and his family to celebrate afterwards at his sister's home in Odon. The mood was ebullient. The judge had acted within political constraints on the court, but had shown considerable restraint in the process. Bill, his lawyers, and witnesses had been allowed remarkably free speech in the courtroom, and from a presumptive sentence of two years in prison and possible four years for felony theft, the outcome had been substantially moderated, first by the jury and then by the judge.

In his closing statement on sentencing, Bill expressed both his personal respect for the judge and his frustration at being confined within legal definitions of the issues. Legal language obscured the truth, and more than anything else confined the jurors' capacity to do justice. The court had been meticulously lawful, but lawfulness itself had confounded justice. And lawfulness in turn had become an instrument of a larger political reality.

The court was subject to even larger forces within the legal system itself. There has always been a tension between judicial hegemony and the Indiana Constitution's provision that the criminal jury is a trier of law. Lay jurors may vindicate defendants in ways that threaten the legal order as judges know it. As guardians of the legal order, judges feel obliged to take matters out of the jury's hands. Thus juries are prohibited from hearing what penalties attach to crimes charged on grounds that judges are the exclusive arbiters of sentence. (In this case, the prosecutor violated his own motion *in limine* by circulating to the jury a newspaper article stating the penalty for felony theft; we took quiet pleasure in this inadvertent blow for jury power.) Juries are also prohibited from considering suppression of evidence, which is understandable since juries should not be biased by hearing potentially excludable evidence at all. But in proscribing juries from hearing evidence of selective prosecution, the Indiana Supreme Court has without other justification excluded juries from considering propriety of prosecutorial discretion.

We know of no case in which a court has dismissed an Indiana prosecution for being unconstitutionally selective. This put our judge in an impossible political bind. Would he be the first judge to tell a prosecutor not merely that evidence was suppressed but that prosecution for this event was impermissible in any form? Would the judge alone assume responsibility for telling an Odon power elite that they had no prosecutorial resource whatsoever? Our inference was that the judge would have been quite willing to let a jury make this decision as an implicit part of a general verdict, but could not possibly take this action upon himself. As the law stood, we could only hope (a) to make the political nature of the prosecution manifest to the audience in court, to the press, and to anyone else who was listening, and (b) to garner implicit sympathy from a judge who ultimately might pass sentence. This we achieved, and from the defendant's perspective justice was done in the process.

The judge closed the other window to jury acquittal all by

himself. To his credit he withheld ruling on whether to allow us a defense of duress long enough for us to present evidence of duress to the jury. The overriding motive behind Bill's taking the sign was to bring the U.S. government's killing of Nicaraguan noncombatants to public and, if necessary, to legal attention. Our concluding witness, Bill's wife, Glenda, for instance, riveted jurors by flipping through a CIA comic book circulated in Nicaragua, showing, for example, how to make a Molotov cocktail to throw at a police station, and by describing worse incidents of terror she had conveyed to Bill after her tour in Nicaragua as a Witness for Peace. Bill's core belief was that his "crime" was a legally justifiable scream of protest against the killing there. The jury got to hear that, as they might not had the judge already ruled that duress was not a defense in this case.

After the defense had rested, the judge told us he had been studying and thinking about the duress defense long and hard. He had concluded that we would have to show imminent threat of serious bodily injury to some particular, identifiable Nicaraguan the evening of Thanksgiving 1986, and this we had not done.

We had not anticipated this issue. We had thought the likely issue would be whether a threat so far geographically removed from Odon could be called "imminent." We had presented our evidence accordingly. It may be fair to suppose that if we had known what mattered to the judge before resting, it would have been virtually impossible to offer the evidence required to establish the defense. The judge did manage to state a ground for denying the defense which would have been hard indeed to overcome in any event. But the fact remains that by giving no advance indications of his thinking, the judge effectively denied us the chance to present what he considered relevant evidence.

This left us and the jury in a difficult position. We knew that after our closing arguments the judge would read the duress statute and instruct the jury that the evidence was legally insufficient to establish the defense. Meanwhile, the thrust of the climax of our evidence was that duress mattered. We could scarcely afford to imply to the jury that this evidence was not important. We alluded to the jury's power to decide for themselves whether the judge's interpretation of the law was correct. We argued that beyond whether duress was established, the defendant's compulsion to save lives overrode any thought or intention of depriving Odon of use, value, or control of the sign.

The judge had heard our objections to his instruction outside the jury's presence. We had no option but to stand mute while the judge told the jury that duress was no defense—to leave the jury with an unrebutted impression that if the judge's instruction mattered enough to be given, it must indicate that evidence of duress was somehow to be ignored. As the jury deliberated, one friend from the audience put it this way: If duress is not a defense and Bill admitted taking the sign, the jury is obliged to convict him.

The judge indicated to us that evidence of duress might still be relevant to whether Bill intended to exert unauthorized control over the sign with intent to deprive Odon of the sign's use or value. Hence he did not explicitly instruct the jury to ignore that evidence entirely. If we had presented the evidence without discussing duress, stressing solely its relevance to whether Bill had requisite criminal intent, the jury might have acquitted him. We have no way of knowing this, of course. But again as Aubert (1959) suggests, a sense of justice would arise from the unencumbered chance that the jury could go either way on this issue. Here the impression was left with Bill and others of us that the judge had biased the jury against considering this issue by waiting until the last minute to snatch the duress defense away.

Again, we are content to believe that the judge had to rule as he did. Had he instructed the jury that they could consider duress and had Bill then been acquitted, the judge could have been accused of allowing Bill's political position to justify a crime. The judge would implicitly have been seen as taking Bill's political side against the Odon power elite. As far as he or we could see, this would have been the first recorded time that an Indiana judge had allowed a jury to apply this defense to any case. That is a long limb for an elected judge to crawl along. The judge also had to see that his instruction would be virtually nonappealable, especially if he kept any sentence moderate. His denial of the defense was consistent with broad precedent that duress was to be narrowly construed. He had, in other words, a lot to gain and virtually nothing to lose by his instruction, especially when he had allowed our evidence to be heard. Here, then, was a politically moderate, personally accommodating action constrained by an inherent injustice of legal realities, which led Bill in his sentencing statement to declare personal respect for the judge but contempt for the system in which he operated. After sentencing, Bill said, "Given the parameters of justice that we have, I think it was a fair trial" (Breeden 1987).

One other procedural nicety rankled Bill and Glenda particularly. The prosecution moved for separation of witnesses. If Glenda, family, and friends wanted to testify on Bill's behalf, they had to leave the court. Since Bill was our lead-off witness, they missed hearing him. From the legal point of view, this kept witnesses from fabricating consistency with others—Bill especially. Bill broke in and asked the judge for permission to speak on his own behalf. He decried this failure of an open court, and of the chance to enjoy the company of close companions during his trial. From his perspective, he and his witnesses were committed to volunteering the truth. They wanted passionately for the jury to understand the case as they knew it. In the world of mutual trust in which Bill and his friends operated, openness promoted truth. It would for instance allow Glenda to hear where Bill missed a point, so that she could fill in information to the jury. In this view, truth-telling rested on cooperation. For his part, Bill offered to be the state's first witness, to admit what they could not prove by themselves—that he himself had taken the sign—so that the jury could begin with full knowledge of the act in question. To Bill, the adversary nature of the proceedings promoted concealment and obfuscation.

The irony was that it became clear that at least some of the state's witnesses, who were talking freely together while they waited to testify, were changing testimony from that given in their depositions, in order to give a false impression that John Myers and Joe Haskins had carefully obtained Town Board members' permission before wiring up the sign that Bill took down. The sad part was that current town employees appeared frightened or timid enough to keep changing their stories to cover for John and Joe. We argued to the jury that Odon was a place where private citizens could play around with others' property—notably to wire up the John Poindexter sign—so that unless the town was engaging in wrongful political discrimination, Bill had as much authority to borrow the sign as did anyone else. The prosecutor responded that Bill was no longer an Odon citizen. Had the jury accepted our argument and acquitted Bill entirely, they would implicitly have had to find reason to believe that a number of prominent Odonians had lied on the witness stand. The adversary trial model implies that for a defendant who admits the act in question to be vindicated, someone on the state's side has to be discredited. Bill did not want to get personal, but figured he had no alternative but to let his lawyers attack the integrity of those who concertedly attacked his. Built on and promoting personal distrust, the adversary trial model seemed to promote fabrication of testimony more on the side seeking separation of witnesses. The formal appearance of integrity ran against the will to be candid.

One of the appeals of mediation mechanisms is the open give-and-take they allow (Witty 1980). Overall, it was the structural impediments to give-and-take which made the trial process unjust to the defendant. Witnesses are to speak without speaking to one another. The judge hears arguments, then rules without giving counsel a chance to gather further evidence to rebut premises of the ruling. Jurors cannot openly discuss the unfolding case with one another, let alone with the parties to the case, and thus their concerns and questions cannot be addressed as they arise. Jurors are not obliged to explain their verdict even after it is rendered. In all these instances, persons involved in telling and unraveling the story are left guessing rather than knowing what is on each other's minds. Give-and-take was the general yardstick by which Bill gauged the justice of the course of his conflict. Crucial to appreciating the defendant's sense of justice is the recognition that for the defendant the trial is a distraction from ongoing life in the outside community. Unless the defendant faces interminable incarceration, the defendant's overriding concern is likely to be maintaining or building relations with family, friends, and neighbors. So when Bill gathered with family and friends, conversation about the case was apt to deal with events outside the court—events not in the Daviess County Seat of Washington, but ten miles away in and around Odon. Two issues were of paramount concern: (1) how much opposition to the Reagan Central American policy was appearing in the locale, and (2) how much courage to resist the Republican power elite in Odon was being demonstrated.

One event alone made the trial worthwhile for Bill. Two sisters of his who live in the Odon area had thought he had gone berserk when he abandoned his parish ministry for peacemaking at the close of the seventies. Now in the street sign case they finally understood what Bill stood for; they wrote a letter to John Myer's paper proclaiming their newfound pride in their brother. (Publication of this and Bill's letters was another sign of justice, for before the case Bill's attempts to publish letters in the *Odon Journal* had been fruitless.) Together the larger Breeden clan discovered a mutual appreciation of moral commitment and political independence that they had shared without realizing it. As much as anything else parties after conviction and sentencing were a celebration of that unity. In one form or another, criminal defendants are bound to face the existential issue of whether there is justice in the larger world. Social solidarity of the Breeden sort offers an affirmative answer, and courtroom dynamics assume secondary importance.

The peace circle in front of the courthouse before sentencing graphically demonstrated the multiple levels on which social solidarity can occur. The circle included more than a hundred people. Bill was flanked by his family. Ministers from Bill's Disciples of Christ Church had come from the region. So had some of their family members. Representatives of Congregations for Peace came from Bloomington. Friends and acquaintances, some of whom had never engaged in protest, came unobtrusively from other parts of the state to join the circle.

As people around the circle made personal declarations, one veteran from a nearby town angrily interrupted several times, saying that none of us had fought for or defended the United States of America. Afterwards he was approached by a World War II widow and a Vietnam veteran from the circle.

He was heard to say that he had learned something, that the experience had given him something to think about. This story was repeated many times after the sentencing as a sign that the case had contributed to justice.

Dicky Ray Poindexter was an object of constant attention. He was seen as the real complainant, the real force behind the prosecution despite the legal fiction that the Town of Odon was the victim. Initially, in his deposition, he disclaimed any responsibility for the prosecution, flatly denying the early calls to them that the former prosecutor and sheriff had said he made (but he acknowledged the calls at the trial seven months later, which we took as one bit of justice). During Old Settlers' celebrations in Odon in mid-August, he threw a friend of Bill's against a wall and knocked a video camera out of his hands because the friend was wearing a John Poindexter shirt with a ban sign on it. We introduced the video tape of the assault at the September hearing on selective prosecution, where Dicky Ray failed to honor our subpoena. We felt that some justice had been done when we learned that the judge had quietly fined Dicky Ray $200 for failure to appear, and when Dicky Ray did honor our subpoena to the trial, wearing a knee brace backwards outside his trousers so that the hip pad flapped against his thigh as he made his way to the stand. We felt some justice when on the stand at trial Dicky Ray objected to Communist literature sitting on the rail of the witness stand, which Guy Loftman then had him read aloud; it turned out to be a message of appreciation to the Sheriff's Department for professional handling of Bill's arrest, with wishes for peace and joy in the December 1986 holiday season. During jury deliberations, Guy Loftman drew some sense of justice from chatting with Dicky Ray and having Dicky Ray invite him over for a steak barbecue. And after the sentencing, we drew some satisfaction from a press report that Dicky Ray thought the sentence had been fair. Mind you, there remains among the Breedens considerable anger at what is perceived as Dicky Ray's highhandedness and arrogance. It is recognized that Dicky Ray is as unlikely to change political stance as Bill is to change his. But through a continuing process of pushing and conciliation, it is perceived that Dicky Ray is accommodating to freer political expression in his hometown, and in that give-and-take lies justice.

John Poindexter returned to Odon on Old Settler's Day, August 18, 1987, to lead the parade, to speak, and to have his street dedicated. About fifteen of us marched and sang for peace in that parade, and drew a sense of justice from the fact that we were more cheered than booed along the parade route. Bill's high school history teacher made up protest shirts (as had Bill himself to raise funds to pay trial expenses—notably for depositions), which he sold along the parade route. This was taken as a sign of evolving justice in Odon. Fear of dissent was dissipating; tolerance of dissent was increasing.

Signs of justice can appear in unforeseen places. When Bill was booked in to serve the three days in jail, he was put in a holding cell because the jail was overflowing. The next morning, the sheriff told Bill that he had had to lock down one cell block where a couple of people were charged with murder, and where unrest ran high. The sheriff asked Bill whether he would be willing to go into the cell block to help make peace there. Bill readily agreed. Here was an acknowledgment of Bill's worth from within the system that had jailed him, and it made Bill feel good.

The judge loves to play golf. Bill used to play golf regularly during his days as a parish minister. He mentioned during trial testimony that he might like to play a round with the judge after the trial was over. Perhaps he will, and if so, Bill will have drawn one more sign of reconciliation and accommodation—one more sign of justice—out of the trial process.

WHAT IS JUSTICE?

Connie Loftman, who manages her husband Guy's law practice, observed that Bill is the most politically aware criminal defendant they have encountered, but that his perceptions are remarkably similar to those of other clients. Either defendants are so beaten down by prior experience as to be cynical, or else constantly are taken aback at the sense that the legal process frustrates getting out the truth and doing justice. Bill's awareness and verbal command enable him to bring out and articulate issues which in other cases are implicit. It probably helped him achieve a greater sense of justice than most defendants, but not to perceive justice in a different way.

The defendant's perception of justice does not rest on a determinate outcome. Justice is not a matter of whether a conviction is valid or a sentence proportionate to the crime charged and the offender's prior record. Sudnow's (1965) classic finding—that defendants are punished according to implicit notions of what the crime charged normally deserves—both makes and misses the point. Even experienced defendants who know their lawyers have bargained for a sentence at the so-called going rate can be left with contempt for the incapacity of the criminal justice system to accommodate to their sense of what they have done and of what they face, with people and circumstances which may never appear in a trial record. It is not merely a matter of a defendant's understanding that the deal obtained is a normal or even better than normal one. In and outside the court, the defendant's sense of justice rests on many levels of awareness that what matters to the defendant affects outcomes which extend far beyond the jail door, both in time and in space.

My inference from Bill's experience is that a sense of justice is the opposite of a sense of violence. Responsiveness as described in chapter 2 gives rise to a sense of justice; unresponsiveness gives rise to a sense of injustice. Bill, for instance, felt injustice at not being privy to how the judge and jury were thinking as they decided on rulings and a verdict—for having no sense of what mattered to them while he still had a chance to respond, as one might have to a mediator. It seemed unjust that the judge had already written out his sentence before the hearing, but then, too, in retrospect the sentence seemed to be influenced by sympathy for Bill's integrity, and in this and other respects the sentence and the earlier trial process seemed fair. Beyond this, justice appeared to have been done by the trial because of outside interaction and accommodation the trial helped generate in and around Odon.

As Guy, Bill, and I sat down after Bill's arrest on December 10, 1986, we could not possibly have assessed how any of us would evaluate three days in jail, eighty hours of community service, and a year's probation on a conviction for conversion, imposed nearly a year later. The verdict and sentence were meaningless without intervening, interweaving, emerging contexts. The meaning of "vindication," which Bill claimed after sentencing, was elusive during the process. At various points when it seemed possible that the state might move to dismiss, Bill even expressed a sense of injustice at the thought that after all our investment in preparing the case he might not have his hearing before the jury. Delays in getting the case resolved frustrated him, but the thought that the case might end without a proper airing of issues almost seemed more frustrating still. He wanted his sundry communities to understand what the case meant to him, with no prior sense of what the communities ought to do in response. The more surprising the response, as in his sisters' letter to the editor, the greater his sense that his case was being heard in a just world, where his presence in the case had a creative rather than a determinate impact on others, where it opened new avenues to accommodation even with opposing parties rather than hardening anyone's previous position.

Justice occurs in and around the trial process in defiance of the formal common-law ideology of criminal due process. Formal bows to impartiality and purity of evidence weighed invite political biases to operate without acknowledgment, and preclude defendants from a sense that reality as they know it influences rulings, verdicts, and sentences, and thereby influences their relationships outside the court. This is an extension of what Christie (1977) means by victims and offenders sharing ownership of disputes.

Many reasons have been given for preferring the give-and-take of mediation, especially with open community participation, to an adversary adjudication process. Now this consideration can be added: The open give-and-take which amounts to corruption or defiance of the adjudication process is for criminal defendants the essence of feeling justly treated. Of course, it is unjust for one party's influence to determine the outcome despite conflicting interests. But justice lies in incorporating and cutting across conflicting interests rather than in trying to purge them from the trial process. Justice entails not blindness, but looking at the world through compound lenses.

The ideological shortcoming of justice as formally conceived in the courts has its corollary in social scientific inquiry. Justice is more than correspondence between sanction on the one hand and offense and prior record on the other. To be sure, justice occurs in the variance unexplained by the additive effects of characteristics of a party of a decision-maker. Justice is exogenous to standard assessments of the justice of verdicts or sentences (Aubert 1959; Pepinsky 1982b, 1976a:44–56), and yet justice is more than random noise. Justice has substance. Justice can be described as interaction and accommodation among persons who are affected by trials, including, for instance, family members given no formal role in the criminal justice process (Smykla 1987). It can be described as novel responses to inherently unique junctures of human experience and needs.

Insofar as justice occurs, it relieves participants of violent tension and encourages further responsiveness, as in respect for criminal justice officials, victims, and the interests they represent. The ideologically pure and unresponsive trial process can be expected to antagonize and generate resistance in criminal defendants. Inasmuch as theories of special deterrance, rehabilitation, and contrition all rest on a defendant's perception that sanctions are fairly imposed, strict formal or substantive rationality can be expected on several grounds to make criminal defendants go bad. In practice, the ideology of justice as blind due process becomes justice denied; and more to be regretted, from this perspective one cannot see real justice where defendants experience it.

We can now see that responsiveness and violence follow their own courses at individual and societal levels, and that the substance of responsiveness, which includes one's sense of justice, takes its clearest form at the individual level. The question remains whether responsiveness and violence, and the links between individual and societal levels, can be given more precise and generalizable form.

5

The Geometric Form of Violence and Democracy

My Norwegian experience was very useful in giving form and substance to the antithesis of violence. "Responsiveness" is a useful concept; as the last two chapters have shown, patterns of responsiveness and violence can be delineated at societal and individual levels. Still, the concepts of responsiveness and violence as thus far delineated are rough concepts. It is hard to know how to operationalize them to make testable the central proposition that responsiveness eases human tension while violence increases tension. To give responsiveness and violence more precise form, I have found it necessary to explore ideas and concepts beyond the Norwegian experience and beyond chaos theory.

If the theory of responsiveness as thus far proposed holds true, then responsiveness and violence ought to take particular forms which remain constant from individual to societal levels. The case study of the Breeden trial in chapter 4 indicates that these forms ought to be more clearly discernible at the level of individual interaction. Ultimately, in a human dyad, one should be able to describe the form of what one might well call a molecule of human responsiveness. The description of that molecule is the central topic of this chapter. The analysis goes on to describe the form by which a molecule of responsiveness at the individual level interacts with responsiveness and violence at societal levels.

As I now move beyond Norwegian usage, I no longer feel constrained to use the Norwegian term for the opposite of violence. "Responsiveness" is not a familiar concept to English-speaking readers. For an audience more general than

Scandinavians, I would prefer to find an appropriate equivalent which connotes "responsiveness" and yet seems familiar across linguistic boundaries. From now on in this book, I choose to call "responsiveness" "democracy."

By "democracy" I mean the principle that those more affected by a decision are more qualified to make it. "Democracy" may be contrasted, for example, to "meritocracy," the principle that those who have earned some credential are better qualified than others to make certain categories of decisions. A meritocrat might believe that legislators, doctors, the religious faithful, or scientists who study when life begins are more qualified than a mother to be the ultimate judge of whether to abort a fetus. As I define it, "democracy" would require that in lieu of the fetus's speaking for itself, the mother is best equipped to be the ultimate earthly judge of the issue. In her, violence in this particular situation is more likely than anywhere else to be tempered by compassion and restraint. Put the other way around, as I believe it works, societal pro-choice consensus indicates that the members of that social system live in a climate of less violence than those in a pro-life state. "Democracy," or "responsiveness" as it has been described in preceding chapters, rests on the perception that actors' objectives are constantly open to negotiation with the persons affected by the action. "Violence" begins where "democracy" ends, as when a rapist decides to go ahead with intercourse, or a purse-snatcher decides to give the victim no chance to resist, or an owner refuses to renegotiate wages with workers, or rule by a head of a nation or household is established.

THE MODEL

A molecule of democracy

What form might the enjoyment of peace or the suffering of tension take? I think Buckminster Fuller (1975, 1979) has given us a clue. He asks us to consider the geometry nature assumes where synergy—the creation of life and other matter-energy—takes place, as against where the Newtonian law of entropy—disorder and death—reigns. The assumption is that in this world, and indeed throughout the universe (Capra 1977), a law of conservation of matter-energy prevails. Wherever matter or organized energy dissipates through entropy, matter-energy organizes itself somewhere else. We know this force of "synergy," for instance, as homeostasis which holds living organisms together. Students of "chaos" refer to this appearance of order in the midst of chaos as "strange attraction," as we saw in chapter 3.

Today we know the form of the elemental molecule of life itself—the double helix of the DNA molecule. Fuller takes the

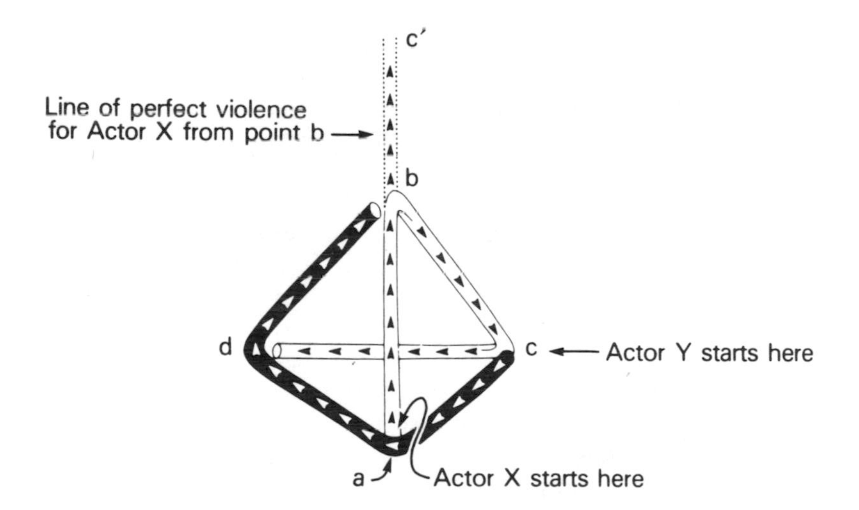

Diagram 3: Tetrahedronal Interaction

Actor X first anticipates where actor Y will be, and moves *in order to* be there first (line ab, "in order to" motive), while Actor Y moves to where Actor X has been (line ca, "because" motive).
The sequence of motives for Actors X and Y respectively is:
X = in order to, line ab / because, line bc / because, line cd
Y = because, line ca / in order to, line ad / because, line db

Perfect violence is to forge straight ahead, for example for Actor X to move from point b to point c′.

double helix to be the form of synergy wherever it occurs: for example, in the resonant arrangement of molecules in a quartz crystal. As shown in Diagram 3, if the helixes are permitted to touch end-to-end, the simplest double helix takes the form of the tetrahedron. Fuller is led to suppose that the tetrahedron is the form dyadic units of matter-energy assume wherever synergy occurs.

The tetrahedron is shown in Diagram 3 as the interaction of two actors, as explained further in the paragraphs that follow. The diagram, with an accompanying description, shows what tetrahedronal interaction looks like.

The course of human interaction is a course of matter-energy—the material form action takes, and the corresponding spirit in which the action is taken. I notice that if one models the most basic unit of dyadic human interaction as a tetrahedron, the interaction takes just the form of give-and-take my Norwegian experience had led me to believe actors feel when they find themselves in the midst of peace, as opposed to violence.

Fuller offers an explanation for our sense of violence. We presumably love life and fear death. Tetrahedronal human interaction literally resonates with the flow of homeostatic life energy in our bodies. It is common to speak of this resonance as being in harmony with nature. When we harmonize, we feel secure, at peace, we enjoy pleasure (French 1985; Harris 1985). When we or others in our interaction pursue an objective compulsively, we feel a threat to our survival. We feel tension, we feel stress, we feel impending death. The violent tension makes people conservative, ritual becomes hardened. People try harder to stay in place and to put others in theirs. People's objectives become rigid; discipline becomes the order of the day. Lines straighten away from tetrahedronal interaction, both personally and organizationally. This is to say that the opposite of synergy happens—entropy, or heat. Violence is social entropy, social heat.

Where a victim of violence manages to engage in tetrahedronal interaction, the tension eases, the victim gains a sense of safety and is better able to forgive and forget. In the physics of this alternative, entropy is absorbed into synergy, something like the not-yet-understood ability of the firefly to give off cool light. The force of synergy explains why music and art have been such strong foundations for community organizing (see, e.g., Craigmillar Festival Society 1987), and why harmony with nature in literature and art was such a prominent part of the Romantic movement's protest against Napoleonic despotism, with "music making" a major metaphor for peaceful change. Notice that tetrahedronal interaction does not presuppose intersubjectivity. No two people need to be in the same place at the same time, or to have a common understanding of what is happening, for synergy to occur. Rather, each actor must constantly shift direction in response to a perception of where the other actor has headed or will head. Compassionate people orbit around one another in double helixes. Since physicists have established that no two particles can occupy the same space, we can probably do without an epistemology which presupposes intersubjectivity. The logic of the tetrahedron suggests that people can and do still move together, each actor retaining her or his individuality. In fact, trying to put people in the same position or to make people head the same direction is violence itself.

As people in tetrahedronal interaction move, they alternate between responding to one another's "because" motives and "in order to" motives (Schütz 1970). That is, they alternatively draw inferences as to each other's objectives from encountering where the other has been or anticipating where the other will be, and changing their own course accordingly.

In principle, the model ought to be subject to empirical test. People could, for instance, review videos of their own or others' interaction. They could encode their inferences as to whether each act anticipated a reaction or reacted to a prior action. They could in many ways encode their inferences as to how actors or observers felt as the interaction progressed, for instance using a semantic differential. The model predicts that interaction would be felt to be most satisfying and least tense when actors were trying to move the interaction forward rather than back on itself, were changing objectives in the process, and were alternating between reacting to past behavior and anticipating future behavior in the sequence the tetrahedron depicts. As investigators became experienced at evaluating the violence or synergy or sets of interaction, they could test whether people who had the greater proportion of synergetic interactions, as groups or societies, also had lower indices of collective violence, such as lower incarceration rates or less entanglement in wars. The sense of violence in the situation is predicted to correspond to the variance from tetrahedronal interaction, as formulated in Diagram 3.

Were the model to stand empirical test, people could begin to describe more fully how synergetic interaction works and is arranged, so that synergetic interaction could be more widely achieved. Were success achieved and the societal level of violent tension reduced, the propensity of people to regard and punish parties to conflict as criminals would be reduced. Crime and the violence in which it is embedded would diminish. A science of peacemaking would supplant a science of meeting violence with violence.

Diagram 3, then, represents an elaborate explanation of how violence and peacemaking occur, and implies a number of propositions about how to dampen violence. It describes violence not as acts but as a relationship between actors' motives. Much of this description has been in micro-terms. It remains to be shown how the model crosses social system levels.

A vatful of violence

Diagram 4 extends the model to crossing social system levels, using a metaphor I call the Schlegel Vat of Violence. (A prolonged talk in a continuing debate colleague Kip Schlegel and I have over whether I could possibly mean what I say inspired me to dream up the model. Kip tells me he thinks he understands the model. That worries me a bit.)

In the model the vat is an inverted tetrahedron, open across its top surface, filled with a highly viscous fluid. When the fluid of human interaction is densest, molecules of the fluid assume tetrahedronal form, at which point synergy is maximized, entropy is minimized, and interaction is crystallized and frozen into social stability. The vat is surrounded by a closed container. At the bottom of the container under the tip of the vat is a heat source. The fluid is boiling, bubbles slowly rising to the surface, most of them created at the bottom of the vat, expanding as they rise toward the surface with less pressure on them.

As each gas bubble passes upward in the fluid, molecules of interaction deviate from tetrahedronal form. In extremis inside the bubble, the vaporized molecules ionize and straighten themselves straight upward. Here entropy, social heat and instability, or violence is maximized.

This fluid has a special character. It lives, that is, it reproduces itself. In the process of reproduction, molecules tend to resume tetrahedronal form (which Bateson, 1979, calls

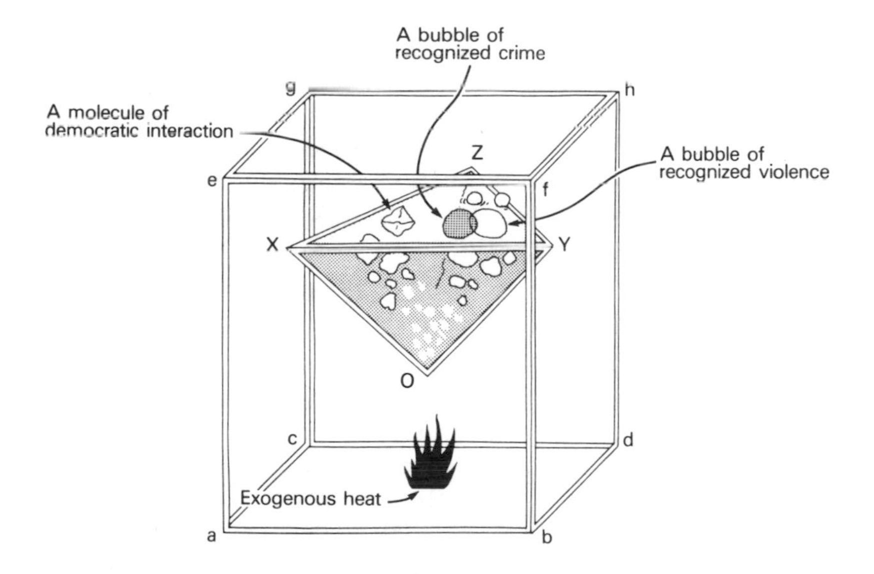

Diagram 4: The Schlegel Vat of Violence

OXYZ = the Schlegel vat of violence = any social system
X = intensity of violence
Y = prevalence of violence
Z = duration of violence
O = bottom of glass structure
abcdefgh = social interaction exogenous to the social system plus formal social control initiatives on the social system

"negentropy"). Bubbles of course collapse when they encounter this coolness. In the same manner crystals are understood to behave, the molecules in the solid state resonate and vibrate in harmonious motion. In this way democratization of human interaction has the potential to cool out violence (see Pepinsky 1985 and 1980:143–67).

That is the metaphor. Here is what the components stand for:

The fluid is a human system of interaction. Any system. In accord with the principle of scaling, the dynamics of the model remain the same whether the system is of interaction in a human dyad, in a family, at a workplace, throughout a nation, in a global economy, anywhere.

Heat anywhere in the container is violence. Reproduction and cooling represent democratization. Gas bubbles represent a noticeable change in the state of interaction in the fluid; they stand for discernible violence and crime. Some bubbles happen to be called crime, others violence. Occasionally a bubble of one is contained in or intertwines with a bubble of the other. The freezing of the fluid is the ultimate state of democratic harmony in some subset of interaction in the social system.

The heat source underneath the vat stands for any organized violence surrounding the entire group, as in workplace violence if the vat represents a family, or legal repression and war if the vat represents a country.

The edges of the cone are axes of three dimensions of violence: its prevalence, its intensity, and its duration. Note that as a bubble rises and expands with decreasing surrounding pressure—as discernible violence or crime persists in the short run—the bubble simultaneously expands along all three axes. That is, the longer discernible violence or crime persists in the short run, the more prevalent, intense, and enduring this body of violence will become. Note, too, that since the external heat source is at the origin of the axes, discernible violence or crime should on average be indifferent with respect to whether it becomes prevalent, intense, or enduring. Still, at lower rates, bubbles will appear and burst throughout the fluid. This is to say that violence or crime may suddenly appear prevalent, intense, or enduring. Where in the social matrix an individual set of violence or crime appears or disappears is indeterminate, unpredictable.

At another level, however, violence and crime are predictable. The more heated the interaction anywhere in the social system, the greater the violence becomes throughout the system. The greater the violence, the more likely violence and crime will become discernible throughout the social system. The more violent the system becomes, the more heat that is released into the container—the social world surrounding the system. This contribution to the exogenous heat feeds back on the fluid. That is, a violent social system spreads violence on surrounding systems, which send part of the violence back. A violent family raises the level of violence in surrounding kinship, neighborhood, workplace structures, which creates a more violent environment for the family. A nation which invades another (as its domestic violence expands) creates enemies. The principle recurs at all levels of social systems.

The line straight up through the vat represents a continuum of power over others. Interaction among the most powerless members of the social system takes place at the bottom of the vat (e.g., closest to external violence, because they are closest to becoming outlaws, to serving on the front lines in war, or to being trapped, as in domestic violence). Interaction among the most powerful members takes place at the top. Note, then, that violence or crime tends first to become discernible at the bottom of the vat, where violent heat is concentrated. Notice that as violence or heat follows the path of least resistance, discernible violence or crime tends to reach the peak of its prevalence, intensity, and duration at the top of the power structure. And the violence or crime at the top is what spills over most directly into the surrounding environment, as in war.

On the other hand, increasing the viscosity or cooling the fluid, through cooperative interaction of fluid molecules by democratization or reproduction, can proceed with least resistance at the coolest part of the social system—at the top of the power structure. Most discernible progress toward solidification of the social system should appear somewhere in the middle (that is, in the middle class, if one exists), where the increasing difficulty of cooling larger bubbles of discernible violence or crime is balanced by the decreasing resistance to solidification in the fluid. But again, where democratization and the giving of life breaks out is in individual instances indeterminate, and wherever democratization or giving life breaks out, it and companion processes throughout the social system contribute to aggregate solidification, and to the net waning of discernible violence or crime throughout the social system. The potential for democratization and life-giving initiatives is always greatest among people at roughly the same level of power over others.

Social systems tend to endure over generations. This being the case, there must be a lot more fluid in the system than bubbles, and the cohesive power in the fluid must be considerably greater than the entropic power of the heat. Which is to say that even if discernible violence or crime blinds us to the underlying reality, democratization must be a much greater force in the immediate lives of the members of the social system than violence. Even people who manifest recurrent, direct violence must generally spend a lot more of their waking hours, throughout their lives, giving themselves to democratic impulses rather than to violent ones. These two sets of impulses, the democratic and the violent, exist in every member of the social system; at issue is the balance between them in prevalence, intensity, and durability. The vector in any subset of human interaction can always shift in either direction—giving way to heat and passing it along, versus resisting violence by devoting marginal energy to increasing democratic resistance. And in so doing, it makes its stochastic contribution to violence or democratization throughout the social system, and indeed throughout the world inside the container.

There is a tradeoff between democratization and reproduction. As we well know, in overpopulated parts of our planet, giving too much life may provide immediate resistance to violence, but overall it increases social viscosity at the expense of generating more heat, in the form of friction among more tightly compressed social molecules. Cooling, by contrast, entails investing more energy in improving the democratic character of life with the people around oneself who have already been born. As democratization enhances one's social security, so childbearing becomes superfluous and burdensome. It strikes me that the optimal position of the middle classes for democratization explains why birthrates tend to drop first at that part of a multiclass social system. Ultimately, democratization rather than reproduction offers the more enduring means to cooling out discernible violence and crime, both within and among social systems.

You can see, too, that trends cannot be projected from trends in the size of any single bubble of discernible crime or violence. Nor do bubbles in the system group by type of crime or violence; any such sample would be biased rather than random to the distribution in the entire system. If, therefore, one noticed a substantial change across a broad array of types of violence and crime, one might reasonably guess that the trend extended to violence throughout the social system. Even a change in prevalence or intensity of violence will have a highly indirect connection to the durability of the violence, especially at the top of the power structure. If, as I think I find through empirical investigation (e.g., of police records, as in Pepinsky 1987a), this model accurately projects how the real world operates, then tracing trends in violence throughout any social system from a dyad to international relations becomes a formidable empirical task indeed. One might better use indirect indicators. For example, if the scale of top-down management of social interaction increases, which in the model amounts to flattening the vat so as to expand the surface area of the fluid, one can project that the boiling process will accelerate, and that discernible violence and crime will on the whole increase. This would for instance be the case for the United States when federal and state government budgets expand faster per capita than local government budgets.

As Wilkins (1984) explains, there is no way to determine whether a model is the same as the real world. Rather, the power of a model rests on how simply and powerfully it works. Here the geometry of the simplest of three-dimensional structures predicts the form violent and peaceful action will take from human dyads to global relations. It predicts how violence and peace emerge and spread. The model transcends partisan political distinctions between criminal and other violence, and predicts how and where in the social structure other violence manifests itself as crime.

The breakthrough for me has been to see violence and crime not as behavior, but as the state of a relationship, of how people orbit around one another. It is not a new insight; you can find it, for instance, in every religious tradition. But it is an insight that has largely been ignored by modern students of crime and violence, who have treated actors as monads.

IMPLICATIONS

Tom Murton (the prototype for Robert Redford's "Brubaker") made Arkansas prisons safe and peaceful during his tenure as corrections commissioner. He told inmates flat out that he knew they really ran the prison. He gave them a high school civics lesson, and offered them a chance to elect their own council, which could make all decisions, subject to Murton's veto (which he never exercised). Rather than trying to abort violence by stripping power from inmates, he empowered them to act openly and democratically (Murton and Hyams 1970). This is the only case in the U.S. I know of where violence in a maximum-security prison has been resolved. It is a paradigm for how to make peace in place of violence according to the model of tetrahedronal interaction.

The model implies that whenever we try to abort violence by dominating violent people, we will find ourselves losing control and encountering greater violence in the aftermath. Rather than doing things *to* offenders, peacemaking requires us to do things *with* offenders and others, or as a Confucian would put it, to lead by example.

The model implies that societies cannot become more peaceful unless and until the members become more democratic in their daily interaction—at home, at work, at school, with friends, in casual encounters. For each of us, no matter how high or low we and those with whom we interact are, the most crucial contributions we can make either to peace or to violence lie in the quality of our daily lives. War and crime cannot be legislated or enforced away. This makes the contribution of any human being to violence or peace supremely important, and evolution from a culture of violence to one of peace fitful and painfully slow.

Researchers on violence and crime are no different. Zuñiga (1975) describes how his U.S. training in experimental social psychology left him utterly unprepared to do research in his Chilean homeland during the Allende period. People were asking him to do research *with* them, to participate in the social experiments he studied, while he had been trained to treat his informants as *subjects,* preferably in a double-blind procedure, where neither subject nor experimenter had any sense of what was going on, let alone a role in constructing the experiment. The method itself is violent. The essential threat violence poses is that the actors in the violent situation go past one another, do not learn from one another how to change their questions, their objectives, their senses of problem, of what is important to know.

Our archives, surveys, and controlled experiments give us but snapshots of interaction between subject and data collector. An anthropologist colleague, Phil Parnell, likens figuring out whether to take people's accounts at face value to peeling away layers of an onion. People protect their privacy with acquaintances, let alone strangers, by telling others what they think is safe and acceptable for them to hear. A facade of peace may easily conceal violent interaction, while loud complaints may cover a situation where conflict is encouraged and democratically managed.

Ultimately for us researchers, our best sense of what heightens and relieves violent tension comes from our experience of our own daily interactions. Here we have a testing ground for our theories of violence and crime, where we can follow the progress of interaction and feel the impact. The measure of how well we understand violence and crime is ultimately how well we understand ourselves. This, the model implies, is what it really means to learn from experience, to be empirical in describing violence and crime.

"Science" means "learning." Scientific convention is that the power of learning is measured by its simplicity, by its parsimony. For instance, a statistic becomes more powerful the more nearly a single item of information (the statistic) describes what everyone in the sample or subsample has in common. Scientific specialization has gotten in the way of parsimony. We expect one set of specialists to explain crime and violence at the level of human dyads (as between victim and offender), another to describe the impact of small-group dynamics on violence (e.g., as subcultures of violence), another to describe the perceptions of people employed to stamp out crime and violence, yet another to describe the economics, or the politics, or the history of international war, and so forth. While it needs to be recognized that not all levels of one's social existence are all at once equally violent and violent in the same way, there is no need to suppose that the principles by which violence and crime occur vary from one social level to the next. There is no need to build independent micro- and macro-theories. How much more we can know if we can take data at the one level and gain understanding of other levels of interaction thereby. How much more we can learn if sociologists, psychologists, political scientists, economists, anthropologists, historians, lawyers, philosophers, theologians can address themselves to a common model, and acknowledge that they are doing no more than describing the same phenomenon in different languages, from different vantage points.

The model addresses the "So what?" question about violence and crime: Crime is inseparable from other violence, and violence costs people their sense of social and self-control, and impairs chances of personal and species survival. Violence is lost opportunity to learn control over one's destiny; violence—forging blindly ahead—is institutionalized ignorance. The model also accounts for the human desire for democracy and peace. And yet the model is not deterministic. It recognizes that individual decisions to pass on violence or to respond democratically are in principal unpredictable, that actors always have peaceful options to violence. The task of a science of violence and crime is to bring those options to our collective awareness.

6

Issues of Citizen Involvement in Policing[9]

People ask four things of the police: (a) that fear and risk of crime be reduced, (b) that disputes be managed, (c) that other services be provided, and (d) that police be accountable to their public. In the U.S. today, particularly in inner cities, the police fail in every respect: fear of crime is rising; disputes are often aggravated rather than resolved; public service, such as looking in on the elderly, is undervalued and ignored; and citizens have little say in how policing is done. U.S. police-community relations are bad.

The problem is that the way the police meet any one of the four demands conflicts with meeting another. For instance, in Indianapolis in the latter sixties and early seventies, police-reported crime went up and down from year to year like a roller coaster. Newspaper accounts indicated that when crime reports dropped, public criticism focused on failures of police accountability, as police failed to record and act on crimes citizens reported to them. So the police would record diligently the next year, only to have a public outcry of alarm over the rise in reported crime. Being accountable aroused fear of crime (described in Selke and Pepinsky 1982; the general problem is discussed in Pepinsky 1980: chap. 7).

If meeting each of the four demands is modeled as a social system of police-citizen interaction, and improvement of police-community relations requires all four models to work at once, then the four systems must be isomorphic. That is, each element of interaction in any one model must be consistent with every interaction in the other three models. Or put in macro-terms, the meeting of each of the four demands must be subsumed within one grand model of how to improve police-community relations by meeting public demands. This approach to constructing criminal justice policies generally is eloquently and cogently described and applied by Wilkins (1984).

The happy experiences in U.S. police-community relations are in upper middle-class communities which, like Menlo Park, California, take pride in their public institutions. When they hire a chief of police with an advanced degree who can require college degrees for all officers and put them in pastel blazers, the police and the community may have a love affair (see Cizankas 1973, as reviewed in Pepinsky 1975a). There the public and the police start with mutual self-confidence. The hard problem most U.S. police face, especially in urban settings, is how to gain public confidence when confidence is lacking. I know of no police department which has succeeded in this effort. The vision I set forth here, then, is utopian. It is a vision of what could be, not an empirical statement of what is. A broad array of data about how members of urban communities interact is drawn upon, much of it more fully described in earlier work of mine, as cited. Here, I try to envision how police-community models could become isomorphic, and to envision the consequences.

MODELING REDUCTION IN FEAR OF CRIME

Fear of crime varies independently of risk of victimization. For example, less-at-risk elderly persons may report high fear of crime, whereas high-risk young men may fear crime less (Merry 1981; Hough and Mayhew 1985). Moreover, police and victim-survey crime trends are not reliable indices of underlying trends in unlawful behavior (Pepinsky 1987a). Thus one cannot infer changes in fear of crime directly from trends in reported crime. On the other hand, the press the Indianapolis police have gotten supports the intuitive notion that when the police either make a wave of arrests (which tend to make offense rates drop; Pepinsky and Jesilow 1985:157–60), or report increased crime, people at all levels of fear tend to become more fearful (Selke and Pepinsky 1982).

It follows that if police can make fewer arrests and report fewer offenses without seeming to withdraw protection from a community, fear of crime at least will not be aggravated by the police. Since the most active police detect and respond to a minuscule portion of the crime in any community (Pepinsky and Jesilow 1985), reductions in enforcement, especially in marginal cases, can occur without increasing people's risk of crime (see empirical estimates for Sheffield policing in Pepinsky 1987a).

Three questions are presented: How can police withdraw enforcement without being perceived as withdrawing protection? How can people reduce their fear of crime with less police enforcement? And how can people reduce their risk of victimization in the process?

Confidence in police rests on citizens' perception that the police are responsive to calls for service, and are visible. Any reduction in enforcement leaves a vacuum in police activity which the police must fill in a way people notice and respect. Respect requires that people who are policed have a sense that they help define what police do (the police respond where they are called). Being noticed requires that people being policed keep account of what their police are doing (Pepinsky 1984; Wilkins 1984). This problem, then, is one of accountability, as discussed two sections below. That is, the police must become more publicly accountable as crime reports and arrests decrease, if public fear of crime is to diminish. But even if accountability goes up, improved police-community relations still require that citizens fear crime less and risk crime less. How can this be done when police are reporting and arresting less?

Fear of crime is an expression of people's sense of social isolation. Within any group, fear of crime may vary among individuals and groups. But over time, fear of crime for everyone in a community should rise and fall together with the degree of isolation people experience from one another (as argued in the chapter on "Societal Rhythms").

Police are few in number and scarcely omnipresent. They are ill equipped to make more than marginal contributions to giving community members a sense of being less isolated from one another. Neighborhood watch programs have been an attempt to address this problem. In neighborhood watch programs, blocks or neighborhoods of residents are called together by the police, given stickers and signs to post, occasionally are encouraged and supported to form patrols or escorts, and ultimately are asked to call their police whenever they suspect anything in their neighborhood is wrong.

The primary problem with neighborhood watch programs is that they play on suspicion and on turning management of neighborhood activity over to outsiders—the police. Either the program becomes a joke where people see that the police are avoiding taking action in order to report less crime, or else the residents and police together succeed in raising the levels of reported crime and arrest (Selke and Pepinsky 1982; Pepinsky 1980: chap. 5). In the latter case, residents discover that the crime problem is becoming worse than it had previously appeared to be. And after the program is in operation, it offers few opportunities for sustained interaction among residents. Those constrained to rely more heavily on having community matters handled by outsiders—here the police—can be expected to find themselves more isolated from other community members in the process (Pepinsky 1980:103–14). Thus it is not surprising that community crime prevention programs like neighborhood watches show increased fear of crime far more often than reduced fear of crime (Rosenbaum 1988:362–63).

While neighborhood residents may build some measure of cooperation in the process of turning people in to the police, the inherent violence of the collective effort builds more violent tension—more of a sense of isolation and vulnerability—than the tension the cooperation relieves (a theory elaborated and empirically grounded in chapter 2 and in Pepinsky 1985). Inevitably, especially in areas of high law enforcement activity, the police will prove to be an instrument for some residents to have others taken away. Some of the residents taken by the police will have kinship ties in the community. Over time factions build within and across kinship networks, and violence escalates (Parnell 1988).

As we have seen in earlier chapters, gaining freedom from isolation and violence entails building community activities which are responsive or democratic instead of violent or oppositional. Admittedly, there is a dearth of empirical literature on the impact of constructive community action on fear and risk of crime. This research occurs on projects aimed specifically at fighting crime, like neighborhood watches (see review in Rosenbaum 1988). As such, these projects are inherently oppositional.

Much of the promising activity of this kind has been organized by women community members and has been founded in the arts. The Craigmillar Festival Society (1987) in Edinburgh, Scotland is a case in point.

Helen Crummy (1987) has lived her life in Edinburgh in the Craigmillar public housing district, which has 22,000 residents. Crummy (1987:1) describes what happened in 1962:

> Asking at the local school for violin lessons for my son the answer was 'It takes us all our time to teach *these* children the three R's, far less music!' Now among my neighbours I knew many talented people—my own father was a Scottish Fiddler! Feeling angry and at the same time helpless, I joined the School Mothers' Club and together with other mothers we came up with the idea of running a local festival. This we felt would not only be a shop window for local talent, it could help combat Craigmillar's undeserved bad press image, and give our children a pride in their area by highlighting our rich heritage of local history.

When the city fathers of Edinburgh spurned requests for help, the Festival Society applied for and received a European Economic Grant on their own initiative, much to the embarrassment of local officials. The Festival has expanded over the years to establishing local enterprises, printing a local newspaper, and encouraging similar projects elsewhere in Scotland. Now that the Society is partially funded by the Edinburgh Town Council, problems of bureaucratization can occur, but on the whole, the Society has contributed substantially to feelings of pride and safety in local community, economically and spiritually as well as socially and politically.

There are numerous promising initiatives in the U.S. as

well. Consider, for instance, what could be done to alleviate the fear and isolation of elderly persons living alone. Residents, perhaps with police assistance, could recruit and help screen people who might share houses or apartments with elderly residents who had extra space and could use help paying the rent or mortgage or taxes. (For steady news on house-sharing projects, see *Shared Housing Quarterly.* Grey Panther President Maggie Kuhn has inspired much of this national effort.) Gang leaders could be approached to furnish visitors and helpers for the elderly in the community, to go shopping with or for the elderly, to go out on walks, to play checkers or talk together. Perhaps young people could also interview elderly residents to construct and publish an oral history of the neighborhood. In such ways, vicious cycles of mistrust between elderly and young residents might be broken. *To be democratic rather than oppositional, community activity needs to be inclusive rather than exclusive, to allow those affected by the activity to participate in shaping the activity.*

Whole cities have pioneered subsidizing democratic neighborhood initiatives, notably for a time in St. Paul, Minnesota (Morris 1986), and to this day in Burlington, Vermont (Clavelle 1986).

The potential for regional development through local democratic initiative is perhaps best shown for the thirty-year history of the community of worker cooperatives, schools, and a bank in the Basque region of Spain, known as Mondragon. Today this community has some 150 enterprises, none of which has ever closed or gone bankrupt. If anything, the community suffers sometimes from too much economic success (Gutierrez-Johnson 1984). It is documented that worker satisfaction and discipline are high (Henk and Logan 1982), indicating that fear and risk of crime have been reduced at least in the workplace.

Grassroots democratization of community life, as we have seen, is the only way to escape the vicious circle in which retaliation for crime escalates levels of violence, including levels of crime. Levels of fear can be presumed to be valid indices of the experience of violence overall, and so should fall as violence generally is replaced by responsive community interaction.

DISPUTE MANAGEMENT

The social isolation of the police has been well documented (among the classic works on the subject—Cain 1973; Clark 1965). When I spent hundreds of hours riding with police in a high-crime area of Minneapolis (Pepinsky 1976b), the police in this "Model City Precinct" were under orders to log "citizen contact" in order to cement police-community relations. The officers logged as they were told, typically by giving the address of a place where they often stopped in for coffee. I could feel the awkwardness most officers would have experienced in walking up to strangers on the street and chatting with them. One officer described to me how police came to socialize exclusively with other police families because of the hostility they felt when socializing with civilians. Another officer, who was in a criminology class of mine where students reported little direct experience with crime, remarked with amazement how warped his view of civilians had become by interacting during patrol exclusively with offenders and whining complainants. Here again is a case where focus on law enforcement promotes isolation and violence. I have repeatedly seen fights between police and citizens escalate out of an ethos of we versus them, where the primary mission of the police is seen by police and citizens alike to be to subdue and dominate offenders. While feelings and experiences may seem to some to be too anecdotal and subjective, it is precisely such feelings and experiences which take the researcher beyond unreliable survey data on crime and fear to a more precise sense of how violence builds in police-citizen interaction.

Law enforcement escalates disputes within the community. Jailed offenders return home with renewed bitterness and stigma. They and their friends and relatives become adversaries of complainants and police.

All in all, dispute management works best when police substitute mediation and conciliation for enforcement, and police can manage this result only if they become familiar with community residents in contexts other than law enforcement. In Christie's (1977) terms, mediation and conciliation amount to giving disputants shared ownership of their disputes—when victim and offender share the power to invent ways to reconcile their differences enough to continue living together. The willingness and capacity of disputants to reconcile their differences in turn rests crucially on their knowledge of one another outside the immediate context of the dispute (Christie 1981). This facilitates the first vital stage of effective mediation—the unrestrained airing of grievances and concerns beyond issues of legal elements of crimes as stated in penal codes (Witty 1980).

What applies to the disputants applies logically to the police as well as to would-be mediators. The capacity of police to reconcile differences in place of law enforcement rests also on their knowledge—in many respects and contexts other than law enforcement—of the people they encounter in the community. This is not just a matter of having enough knowledge to mediate effectively by proposing innovative resolutions to disputes (Witty 1980). It is also a matter of feeling relaxed and able to step outside an authoritarian posture. The more contexts in which police know people in the community, the more warmly and impartially disposed toward disputants they will feel, the less threatened they will feel, the more willing to forgo formalities and resolve matters informally they will be. To paraphrase W. I. Thomas's famous dictum about defining situations: Police who know people well enough to try mediating disputes with confidence will really succeed in dispute management.

There is a chicken-and-egg problem. The police cannot know citizens well in a community where members are isolated from one another. Community members have to act independently of police initiative to overcome their own fear and violence. The police can scarcely relieve fear and violence by themselves, not even on their own initiative alone. But in a community where people are otherwise willing to build their own social structure of peaceful relations, the police may facilitate the process by accommodating to the change and involving themselves in community activities other than law enforcement. For instance, they might join young residents in visiting and helping elderly neighbors. In this way they would get to know youth outside the context of decisions whether to arrest, and elderly outside the context of whether to take action on complaints of crime. All in all, police for their part need to be prepared to perform service roles in lieu of attention to law enforcement. In this context police involvement in informal dispute management can grow.

PERFORMING SERVICES

It is commonly accepted and never debated that the bulk of what police do is not law enforcement, but falls loosely under the heading of service. "Service" can mean many things—popular examples include getting cats out of trees, finding lost children, and helping people who have locked themselves out of their homes or cars. British police constables were once renowned for providing all manner of useful services. Now that role is said to be largely a thing of the past, although one hears stories of bobbies feeding the cat while residents are away on vacation (Cain 1973).

Police do tend to downplay and denigrate their service functions, except in the case of some flashy avant-garde police programs in prosperous residential suburbs (Pepinsky 1975b). As one high British police administrator put it to me, it is hard to give as much emphasis to other activities as to law enforcement when the formal police accounting system gives preeminence to counting offenses or arrests. If the police officer knows that for advancement numbers of arrests in the personnel file will weigh far more heavily than all the officer's other good deeds, the officer is not so likely to go out of the way to do other good deeds.

The problem of encouraging police to substitute other service activity for law enforcement thus becomes one of identifying, recognizing, and rewarding the other activities as much if not more than law enforcement. This, essentially, is a problem of police accountability.

ACCOUNTABILITY

With the police, as with all actors, two basic models of accountability exist: the model of accountability of the actor to superiors, and the model of accountability to peers or subordinates.

The first model has been variously referred to—for instance by Brock-Utne (1985) as a "system of domination," or by McNamara (1967) with specific regard to police as a "quasi-military organization." This system actually generates more unaccounted-for, unbounded police discretion than it controls, promotes a narrow orientation toward law enforcement, and isolates police from their communities (Pepinsky 1984).

The other model is what was called "democracy" in the last chapter, where people are presumed qualified to find out about and evaluate decisions not in proportion to credentials ("meritocracy"), political position ("autocracy"), or birthright ("aristocracy"), but in proportion to how much they are affected by the decisions ("democracy" = "rule by the people"). In the case of police and other officials, democracy amounts to public accountability in the truest sense.

It is easy to imagine how democratic police accountability might be arranged. The key would be to give the people policed control over the flow of information into and out of police personnel files, as an open-ended process of seeking consensus with police officers and administrators. To summarize the process more elaborately developed and described in Pepinsky (1984):

Police assigned to a district and their supervisors would arrange a series of town meetings under the auspices of various community members and groups. At these meetings police and community members in attendance would brainstorm about activities police could perform in the community. Over the course of several meetings consensus could be sought on activities to be specially valued and rewarded. These might include law enforcement. The hope would be that over time people would think more broadly and imaginatively about things police might well do. Police/community priorities could be added to or revised at any time as community meetings continued to be held at convenient intervals.

Once priorities are determined, community members and police together would devise a system by which community members could gather evidence of police performance. If for instance a priority was given to courtesy visits to elderly residents living alone, community members and police might design a survey instrument which police would leave with the elderly residents they visited, to be collected and coded by a committee of community residents.

Ideally, police administrators would agree that whenever an activity had been given priority, as community evaluations of service became available, the data and evaluations provided by the community would be placed in the personnel files of the officers concerned. Ideally, too, no other evaluation data, other than records of complaints and disciplinary proceedings, would appear in the file. If the community were gathering data on arrests, counts of police arrests could go into personnel files, otherwise not. As a rule police would not count arrests or offenses at all, but would keep data from offense and arrest reports only in individual case files.

The larger legal duties imposed on police not to violate anyone's civil rights would act as a natural check on community members' proposals to use the police to conduct vendettas, chase out outsiders, or the like. Community/police proposals for priorities and evaluation procedures would have to be reviewed and concurred in by higher-level police administrators responsible for relations across police districts. With this requirement of police/community consensus, the system could give community members a real chance to guide and diversify the conduct of their police. Police and their community members would become allies in demonstrating that what the community wanted was appreciated in higher police circles, as by use of personnel files in promotional examinations.

Of course community members' participation in this system of accountability would be enhanced in a community where members had otherwise mobilized to work together on a range of projects having nothing direct to do with law enforcement. Here, too, more options for police to be of service would present themselves, both to community members and to their police.

CONCLUSION

The principles by which to enhance public safety, manage disputes to everyone's satisfaction, reward police for diversified service, and hold police accountable to their citizenry all are isomorphic. In each of the four subsystems success rests most heavily on broadening the experience and skill of community members in working together to democratize all manner of community life, and then on the willingness of police to help community members take care of their own disputes and manage their lives without isolation either from the police or from other community members.

This isomorphism, built as it must be on empowerment of members of a community to manage their affairs democratically, is an instance of what was referred to in the preceding chapter as "synergy" (see also Ivanicka 1988a, 1988b). As community members gain power to interact and cooperate to meet one another's needs and concerns, this flow of current generates energy in the form of compassion, essentially the bond of interpersonal cohesion, which alone inhibits violence, including crime and retaliation for crime. This is the only glue which can bind police and citizens in conflict management as against escalation of violence.

We are a long way from implementing this model in U.S. inner cities. But that should not keep us criminologists from imagining what citizen involvement in policing might become in an ideal world. Without such visions, there can be no progress. Organizational change is harder to achieve and no less significant than change at the level of individual interaction, which is the focus of the next chapter.

7

Speaking Freely with Children as a Path to Peace

THE SIGNIFICANCE OF VIOLENT CHILDREARING

The ultimate empirical test of human progress within our own lifetime is how our children turn out. Small wonder that child development is a focal concern of social science, theology, medicine, and law, and, yes, of criminology itself. Adolescence has been created in a mobile, industrialized, urbanized world. Adolescence is that great period of uncertainty when adults search anxiously for signs of how the heirs to their cultural legacy are turning out and when adolescents search eagerly for badges of respectable adulthood. Crime and other metaphors become vehicles for adult expression of concern over whether children are carrying forward the right values.

This concern, I think, is quite valid. As a college teacher in the heartland of the U.S., I struggle with the institutionalized brainwashing in militarism that U.S. students are expected to know and believe. My students are unlikely to be arrested and killed for political action, as Korean or Chinese students are, for example. My students are more likely by far to be arrested for public intoxication or underage consumption of alcohol. Nevertheless, they are often desperate to show that they can be good soldiers, not only in the military, but in law enforcement or in prosecution or correctional supervision. The wider fear is that since all U.S. citizens know that every individual rises or falls on his or her own merits, one dares not question authority and one must succeed in the competitive struggle to conform best to the established order. Tocqueville (1956

[1840]: 303–04) succinctly described this as the kind of despotism democratic nations (as epitomized by the U.S.) had to fear, a despotism which

> covers the surface of society with a network of small, complicated rules, minute and uniform, through which the most original minds and the most energetic characters cannot penetrate, to rise above the crowd. The will of man is not shattered, but softened, bent, and guided; men are seldom forced by it to act, but they are constantly restrained from acting: such a power does not destroy, but it prevents existence; it does not tyrannize, but it compresses, enervates, and stupifies a people, till each nation is reduced to nothing better than a pack of animals, of which the government is the shepherd.

What Tocqueville failed to recognize is that this kind of discipline, later called fascism or totalitarianism, makes this kind of "democratic nation" go to war, against foreign and domestic enemies. (See Zinn's (1980) excellent account of the history of U.S. violence.) We have repeatedly elected as President military leaders and heroes—Washington, Jackson, William Henry Harrison, Taylor, Grant, Teddy Roosevelt, Eisenhower, Kennedy, and now Bush—and have even elected a mythical military hero—Indian fighter Ronald Reagan—President by modern landslides. We have fought wars against foreign enemies continually since the founding of the Republic, and since World War II have not bothered to fight any of our continuing, more or less covert, wars constitutionally by having Congress declare war (Moyers 1988). These wars have considerably tightened alliances of convenience between U.S. politicians and agents on the one hand and international illicit drug cartels on the other, to buy weapons for U.S. covert operations (Chambliss 1988). Meanwhile, on the domestic front, we are in a boom for private construction and management of prisons, even as one in twelve black men in his twenties already sits each day in the nation's jails and prisons, with more than twice as many on probation and parole (Pepinsky 1987b). California officially projects having as high an incarceration rate as any I have heard—800 per 100,000 population—by the turn of the century. Only Argentina during the junta's recent reign of terror managed that level before, as far as I know.

The supreme good in this political culture is for everyone to know her or his place. We start on children early, telling them they are dumb, ignorant, and irresponsible. It is the premise of a brilliant new delinquency and juvenile justice text by Regoli and Hewitt (1991) that this training in obedience and degradation is the primary cause of delinquency. Our first lesson in the ideology that the mightier are righter is the difference between being small and young and being big and old. (Much later we learn to our chagrin that being too old is also a liability.) As we mature we learn refinements on how to separate those who give orders and do what suits them, and those who do what they are told. We learn that men tell women what to do. We learn that the mightiest men are those in charge of the U.S. government, without whose grace we would be poor, backward, or dead. We learn that boys can occupy more space than girls (Brock-Utne 1985: chap. 3). We learn that those who repeat back the adult liturgy best on school tests are mightier than those who rebel. We learn that male conquest of one's opponents on a field of physical encounter makes one mightier than one's fellows (albeit subject to ultimate domination by those who score higher on the tests, who get to order the tough guys into battle). We learn that as we grow big, even as we continue having to take orders, we can find smaller or weaker people to push around in order to have our turn at domination ("Fair's fair, I had to go through it, welcome to the real world, take that").

As long as this force prevails in childrearing, the cycles of violence I have described earlier will continue to recur. Each generation will recapitulate its ancestral violence, out of conviction that maintaining domination and militarization throughout private and public life is the only way to be practical, "human nature being what it is." People wrapped tightly in this culture, as in the U.S., can justly marvel at life in another culture such as Denmark, where schoolchildren will give the Nazi salute to a teacher who tries to order them around, and where prisoners and guards are free of physical fear (Selke 1989). And yet getting from the one culture to the other seems so far to go, so unimaginable, at least to most of my students.

TEACHING CONFIDENCE

There is a quality I call "confidence" which allows people to question authority and get respect. My mother did a series of studies on this phenomenon. She calls it "productive nonconformity"—"productive" because a group recognized its value, "nonconformity" because the action or proposal was independent of group norms (in contrast to "negative conformity," which is simple opposition to norms). She isolates one factor as the major if not crucial contributor to productive nonconformity: having someone of status stand up for the nonconformist expression. She calls this "political sponsorship" (Pepinsky 1961). Similarly, Pines (1979) finds that "superkids"—children so abused they ought to grow up to be a mess but who turn out okay anyway—in all cases have some loving, respecting adult in their lives as they grow up.

I put these two findings together and infer that there is positive feedback between political sponsorship and willingness not to conform. One source of this belief is having seen students at a nearby alternative school, Harmony School, grow over some years. (I teach my seminars there, and have known students and staff there for years.) Many of the students there have been extraordinarily abused. But, in high school especially, students become confident public speakers and highly worldly-aware, mature discussants of social issues. It is particularly remarkable to see quiet, reserved students gain confidence in expressing themselves. The primary effort at the school is to democratize life there, from equal pay for all teachers regardless of seniority, to acceptance of students regardless of ability to pay tuition, to having students and teachers together design curricula, select new students and teachers, formulate school policy, and handle conflict. There are no grades, but among the worldly wisdom that is taught is how to play the game of taking standardized tests, so that a high percentage of graduates become merit finalists. Some students choose to go on to college—often to highly innovative programs—while others develop other interests and at least delay going to college. The school has never bothered to seek state accreditation, but is nationally recognized as a leading educational innovator.

When Harmony students talk to my college classes, they are often asked whether they aren't being sheltered and left unprepared for the real world. Perhaps it is because I went through an ungraded alternative high school myself, but I share the view that the security and democracy of life at Harmony strengthens them for facing the real world and gives them a sense of personal control over their lives which my college students so often seem to lack. Harmony students learn that democratic life is real and practical, and that by investing in democratic friendship with others, they have a safe place to nurse their real-world wounds, restore faith in themselves, and gain the strength to carry on. In a word, they learn confidence.

I take the Harmony experience to be paradigmatic. Children learn to become democratically inclined adults by being democratically treated as children by adults. It is not only that children thereby learn that democracy is possible. They also experience that democracy feels good and builds personal strength and security. They learn that investing in friendship pays off better than investing in wealth and power. They not only can be democratic, they want to be democratic. They feel human synergy, and crave more.

Knowing how to raise children democratically doesn't make it easy. Harmony School itself does an extraordinary job of raising several hundred thousand dollars a year to support a hundred students and staff, where 75 percent of the students are on fee scholarship. The waiting list to get into Harmony is twice the size of the school. In theory, the U.S. could do as Denmark does by constitution: All schools in Denmark are supported by state subsidy according to enrollment and yet all schools are independently organized. All Danish students have the right to go to any school they choose. Harmony costs less per capita than public instruction, and under Danish law the Harmony model would surely proliferate. The trouble in the U.S. is that politicians and an electorate who have grown up in hierarchical schools and families lack the political will to make this change. If that political culture is ever to change, enough children have to grow up in publicly unsubsidized, small-scale democracy for the society to break out of the cycles of violence in which we in the U.S. now live. Change in political culture rests on grassroots change.

Happily, as indicated in the chapter on violence as unresponsiveness, individual decisions whether to treat children democratically are in principle unpredictable. At the level of individual interaction, humans retain free will whether to respond democratically or violently to one another. The odds of democratic response from someone who has learned confidence in democracy may be higher than odds for the abused child, but as Pines's (1979) "superkids" indicate, no one is certain to follow this pattern.

If knowledge is power, as a writer like me presumes, then an awareness of the choices we have as adults should empower us to raise our children more democratically. For many of us adults, our greatest opportunity (and our greatest liability) is to decide how to govern ourselves, with our children, as parents. Parenthood, as Ruddick (1989) cogently argues, is at once an exercise in power and an exercise in humility. Children have wills of their own, and are subject to many worldly forces beyond parental control. To imagine what a parent should be is to confront one's own powerlessness. And yet if a political culture is to be democratized, nothing is more vital than the minute contributions each parent can make to giving children confidence.

PARENTAL DILEMMAS

The following scenarios happened. Anecdotal evidence is often defined as weak. On the other hand, as cultural anthropologists and folklorists well know, the stories we tell are much of the way we learn from one another in day-to-day life. Surveys may report what people do, but stories tell what people mean by what they do. People generally live by stories. Think of how crucial believing children's stories has been to our exploding awareness of child abuse. This is the methodological premise of Regoli and Hewitt's (1989) text on delinquency and juvenile justice. If you want to understand why children hurt and disappoint others, let the children tell you their stories.

But here, as Ruddick does with "maternal thinking," I am concerned with the stories of the parents. After all, as parents we undertake responsibility for our own stories on how parenting should and should not be done, not only for what parents say to each other, but more especially for what stories parents tell the children.

The aggressive scenario

The parents have talked the matter over and agreed. The child shall be in bed by nine o'clock. One parent undertakes to convey the order at 8:30.

"It's time to get ready for bed now. You have to be in bed, lights out, by nine. You have school tomorrow."

"I'm busy."

"What are you doing?"

"I'm coloring."

"Now put that stuff away and go brush your teeth."

"I don't brush my teeth before I change into my jammies."

"Put it away now."

"I have to finish my picture."

"I'm going to give you five minutes to put the stuff away."

Ten minutes:

"What are you doing? What did I tell you?"

"I'm almost done, but I'm starving. Could you fix me a hot chocolate?"

"No I won't, you should have asked earlier. Now it's too late." (Goes and grabs the crayon out of the child's hand, impatiently gathers up the stuff. Child begins to cry.)

"I'm gonna throw up I'm so hungry. I can't go to sleep if I don't have the hot chocolate."

"Shut up and get ready for bed. Drink water in the bathroom."

Child yells, "Nobody tells you when to go to bed. It's no fair you're telling me just 'cause you're bigger. I can go to bed when I want to."

"No you can't."

"Why?"

"Because I'm bigger than you and I get to stay up. That's enough. I don't want to use force, but if you don't march to the bathroom right now, I'll have to spank you."

Child starts screaming things like "you can't make me go!"

Parent shouts, "If you don't shut up and get to the bathroom right now in ten seconds, I'm going to spank you." (And parent does spank child, who runs screaming to the bathroom, slamming the door on the way.)

Parent and child both feel disrespect, shame, frustration, and failure. As I define violence and democracy, this interaction epitomizes violence. The parent has the objective of getting the child to bed by nine o'clock. The child has the objective of defying parental power. The further the interaction continues, with both actors forging straight ahead, the higher the tension mounts in both parties. When violent tension eventually discharges as physical force, it goes toward the point of least resistance—the child, who is smaller and has to "cry uncle" (as Ronald Reagan once demanded of the Sandinistas). And the tension will remain unresolved unless parent and child apologize to one another, forgive one another, and reconcile. If the violence persists, the child may soon learn to obey and be in bed by nine. But the child who learns to accept that might makes right learns if necessary to store the violence up and expend it on safe targets—on the battlefield, on police patrols, in schools, in boardrooms, at home with women and children of our own. That is the social price of achieving compliance. The parent's demands for obedience are meanwhile apt to proliferate out of control, as parent learns that child does not really want to cooperate and has to be forced.

The passive aggressive scenario

Begins like the first scenario, except that the parents haven't agreed on a bedtime. Second parent intervenes after first parent gives child five minutes to put away the coloring material.

The second parent says, "Let the child stay up. Once the child has trouble staying awake tomorrow in school, the child will get to bed early. Nature has a way of taking care of itself."

The first parent backs off, then later angrily tells the second parent not to interfere. "It's not good for parents to give a child conflicting messages."

The parents argue awhile about the merits of letting children go to bed when they are sleepy, and of who is not to contradict whom, and fall asleep emotionally drained and alienated from one another. The child learns to try to sneak out deals with the second parent behind the first parent's back. The first parent feels like the only responsible person who ends up driving the child into the arms and love of the second parent. "I'm tired of being the ogre, and you don't love me."

The violence encompasses three dyads of interaction. For the first parent, every encounter with the child becomes a test of will and power, while for the child, every compromise with the parent becomes a surrender to sheer might. For the second parent, the child becomes a purely manipulative unilateral actor, while the child fears losing control and parental contact. In effect, the second parent gets love for expressing unconcern for what the child does. As between parents, each has a growing conviction that she or he is right and the other parent is wrong, as each tries to subvert the other's negative influence. The parents pretty much stop talking to each other, as each forges straight ahead, and the child goes its own way, ignored at best and chided at worst.

A more democratic scenario

Parents agree to try to salvage their passive aggressive relationship. As friends counsel, if for no other reason than giving parents some time alone with each other at night, child needs to get to bed. Second parent, a great would-be egalitarian, is confronted with not being very egalitarian toward the other parent. Parents agree on 9:00 as a bedtime target, but that delays until 9:30 may be tolerated. Second parent accepts responsibility for helping meet that target.

Second parent initiates interaction with child, while first parent waits to see what happens.

What happens varies from night to night. Second parent, the one who at one stage had spanked the child as in the first scenario, treats 8:30 as time to start playing games with the child. The next hour is the child's if the child wants it. The second parent cajoles the child, sometimes step by step, to prepare for bed. The child bargains for doing some fun things with the parent, like reading stories or singing songs, or playing a few rounds of a board game while eating a last snack. Sometimes the child just wants to get to bed. Sometimes the child wants to be in bed by just after 9:00 and have time to read or color quietly in bed, and on occasion it is sometimes 9:40 before a final hug from both parents and the lights going out. Sometimes the child is too excited or troubled to sleep, and the parents learn to accept, if at times with irritation, occasionally repeated late night cries and visits. Meanwhile, the parents have a lot more time together, and feel a common sense of accomplishment over parent-child interaction.

Paradigmatic of democratic interaction would be the following exchange, expressed in terms of the tetrahedron depicted in Diagram 3 in chapter 5.

Second parent is actor Y, sees child (actor X) bent on coloring a picture, and because of where the child appears to be headed, says, "I'd like you to start getting reading for bed." (That is, parent moves from point c toward point a.)

Child, anticipating that second parent will eventually get anxious if child isn't in bed by 9:15, responds, "How about fixing me a cup of hot chocolate while I finish this picture?" (That is, child moves from point a toward point b.)

Second parent, anticipating that child will want to stay up past 9:00, counters, "All right, if you can finish and get things put away by the time the hot chocolate's done."

Child, responding to parent's original desire to see progress toward bed, concentrates on finishing up the coloring.

Child asks second parent to read a story while the child drinks, because the parent has already proved to be in a good mood by fixing the hot chocolate without much ado.

Second parent responds to the child's compliance with the

parent's wish for the coloring to finish, by agreeing to read a story.

Note that parent and child retain distinct motives throughout the interaction, that one motive continually gives way to another, and that each actor alternately responds to what the other has done, or in anticipation of what the other actor is headed toward doing. This is in an ordinal approximation of the tetrahedronal form hypothesized to be synergetic, to give satisfaction, in place of frustration and tension, for all actors. No single actor is independently responsible for changes in either the actor's own or the other actor's motives. Each change in each actor's motives is an interaction of what both independently seem to be after. If what the parent is after is a function of what the parent wants and the parent's perception of what the child will or has wanted, the resultant motive is a pure interaction term as surely as the green that results from combining blue and yellow is in no measure blue or yellow alone.

This is no utopian family. Conflict recurs and so does negation, but often even then everyone involved repeatedly descends into violence, hurts a parent or child, and lives to regret it. But the point is that a little dose of democracy now and then can substantially relieve whatever violence has accumulated in the home, or, for that matter, outside the home—in freeway gridlock, at work, at school, wherever. It serves in some perhaps small measure to restore family members' faith in themselves, and in their capacity to live securely and fruitfully with one another. Here too experience of democracy builds confidence in parents and children alike. The democratic interaction creates a cool spot for all actors, giving them relief from a violent world, and building their capacity to resist violence rather than passing violence along.

CONCLUSION

A first step toward building democracy is to be aware of the difference between violence and democracy in one's own daily interactions. Nowhere are those interactions more crucial for the possibility of reducing societal violence than between parents and children. The irony is that decisions which appear inconsequential at the micro-level can mean so much for human progress toward peace and democracy at the macro-level.

The three parent-child scenarios as presented above occurred all in the same family over a short period of time. A family, indeed a larger society, always has a dual capacity for violence and democracy. I suspect that a little personal investment in democracy can go a long way toward compensating for the violence all of us humans share.

There is no shortcut to peace. We cannot raise our children undemocratically and expect them to know how to create democratic plans and policies on a grand scale, let alone in workplaces and homes on a small scale.

Earlier I dwelled at length on the causes and consequences of violence. Violence is a common preoccupation of us criminologists. When, as now, I struggle to envision the opposite of violence, I can think of no more meaningful or more vivid setting for it than in adult-child interaction. Ageism is the ultimate barrier to peace. While I can dream far-out dreams of how to reorganize life on a grander scale, as for police, my daily life with children makes change more tangible, more concrete, more manageable. It is a good place to start one's journey toward peace.

8

Conclusion

A Ugandan friend, Karagwa Byanima, once told me that she was glad that I, as an American, was a pacifist, but that if I had seen family, friends, and neighbors die at the hands of an unrecalcitrant, brutal leader, as she had done in Uganda, I could not remain a pacifist. And I might see that in the aftermath, limited use of capital punishment would be required to forestall a bloodbath of vengeance.

So, too, Norwegian friends like Per Ole Johansen have explained to me how Norwegians came to execute twenty-five Nazi war criminals, and of how they had stopped carrying out the death sentences halfway through because enough time had elapsed since the war, and enough blood had been shed, that executions no longer served the cause of peace.

As I struggle with these issues I recognize that in my own life I also act in violent defense of myself and my friends. I don't shed blood, but I do try to march relentlessly through courts or grievance committees, for instance, to force others to relinquish power. Like Gandhi (Sonnleitner 1989), I recognize the principle that violent resistance is sometimes less evil than letting greater violence go unabated. I accept the principle that violent defense is justified *provided that one minimizes the force necessary to make peace.*

Bayley (1976) has described Japanese police as trained in this principle. As I hear sad stories of U.S. police behavior, I cannot help thinking that training in this principle would make a big difference here. Instead, many U.S. police are trained first and foremost to dominate situations, to be "authoritative" (meaning authoritarian). Violence is underplayed by euphemism, so that shooting a gun becomes "using your weapon" and browbeating a person becomes "questioning the subject." One popular set of training films shows people coming toward you, and as they make a suspicious move, the trainee is to decide instantly whether to kill to avoid being killed, and to perhaps shoot an innocent person. Is that a gun or a wallet the person is reaching for? There are no options to shooting or being killed. Trainees are given no cover to hide behind, no opportunity to talk the person down. Then they are given the guns with the most "stopping power" the police can afford, put in cars with the most powerful engines the department can buy, and told to enforce order. This mission is radically different from that of minimizing the force necessary to keep or make peace; it is offensive rather than defensive. Purely defensive violence, then, falls far short of exercises of offensive power.

"The problem after a war is with the victor. He thinks he has just proved that war and violence pay. Who will now teach him a lesson?" (A. J. Muste in 1941, as quoted in Zinn 1980:416). Herein lies my ambivalence about defensive violence. One of my bigger intellectual disappointments has been to follow the course of the Chinese revolution since 1949 (Pepinsky 1982b). The revolutionaries know how to subdue political enemies. They teach their children the virtues of the revolution, and ultimately the children have nowhere to revolt except against their elders, who in turn suppress the youth to save the revolution. Within two generations, traditional dynastic forms of legitimization and of wielding power recur.

U.S. criminologist Bill Chambliss tells me I am blind not to see that the Chinese have made considerable progress through revolution. The only time of mass starvation was during the Great Leap Forward. Epidemics of deadly disease have been abated (although other social diseases such as mass executions of up to 30,000 people around 1984 still take many lives; Tifft 1985). I wonder. Conceding that the overthrow of feudal despots opened the way to democracy and peace, I'm not sure the revolutionaries have been able to stop making war long enough to take enduring advantage of the opening. One lesson is, I think, to be drawn: the grander the scale of the defensive violence, the harder it is to stop the offensive violence in the aftermath. This is the paradox of all wars to end all wars, of all final solutions to problems, of all violence which aims to make things right above and beyond stopping what is wrong.

The greatest challenge of resisting violence is to make democracy in its place. Democracy is not a formal structure, an edifice, or a social or industrial plan—some material form in which people can safely be placed. Democracy is a way of life. Insofar as democracy is lived, it is lived in our everyday mundane activity, and, like walking, it becomes easier the more it is practiced. Democracy begins when the warrior begins to show mercy.

If democracy is not a formal structure, it is no less a real structure. Democracy is a pattern by which actors' motives change in relation to one another. Much as some social scientists would like to get us out of the black box of the human mind into analyzing what people do, much as some criminologists would love to make the punishment fit the crime alone, we act crucially on motives—our own and others' as we perceive them. Without motives, human atrocities that routinely happen around us remain senseless. Without motives, we ourselves become senseless—incapable of any form of rationality which rests on choosing which course of action will get us more of what we want than other options.

With attention to motives, on the other hand, we become capable of sensing violence when it occurs, and of grasping opportunities for relating to others democratically, as though saying to oneself, "Woops! Maybe it's time I tried changing what I want." I often suggest to my students that becoming democratic means investing in friendship rather than investing in wealth and power. People I know who have invested heavily in friendship will never be homeless no matter how destitute, neither by choice nor by force of circumstance. At the micro-level, even if our democratic or defensive initiatives haven't made a dent in the violence all around us, we have remarkable power to make our immediate lives more democratic and hence socially secure.

Small, open, democratic networks give us and our offspring a chance to survive awhile longer. It is cause for optimism that we have supported each other enough not to have exterminated ourselves long ago. One cannot account for population growth simply by fertility. The miracle is that each mating pair manages to produce more than two offspring who live long enough to bear their own children. Human extinction is no further away than a few days of child neglect. I take comfort in the thought that macro-level human forces mirror micro-level limits, that the same hands which cannot beat their own children to death in a routine fit of rage cannot organize pushing the button to set off nuclear war. All in all, human survival is miraculous, our capacity to learn from one another and to shift motives awesome.

Life always hangs in the balance. Long before people split the atom, numbers of people were utterly convinced that the end was in sight. This is a testament to the level of instinctual alarm we feel over violence. Given that we live on the brink, this alarm is a blessing. While violence rages all around us and throughout our own lives, we keep pulling back from the brink only by constantly giving ourselves democratic respite, by giving ourselves and those we relate to a break from violence. We calm ourselves, we cool ourselves down.

I surmise that because the rightful alarm over the violence in which we live reverberates so loudly in our conscious minds we remain largely unconscious of the democracy in which we also engage. We love without thinking about it. We relegate loving to the private or "maternal" world (Ruddick 1989; Brock-Utne 1985), inappropriate for public discourse. We notice our grievances, we dwell on them, we plot to foil our enemies or competitors. If we are sophisticated rather than naive and primitive, we talk with violent authority, and respond to violence in kind. To do otherwise in public discourse, and hence in our thoughts, makes the speaker a sucker, a sissy, or a fool.

It is hard to imagine how much violence we could abate if everyone paid conscious attention to trying to respond to violence with doses of democracy. In theory, there is no limit to the pace of change. And yet it is harder still to imagine everyone agreeing that democracy is worth attending to in the first place. In fact, people may have so much violence stored up in them that they play the would-be democrat for a sucker, or kill the peacemaker. It takes a lot of confidence—a lot of democratic interaction with friends to fall back on—to walk away from a violent response to the broaching of peace, let alone to turn the other cheek.

People may gain and spread confidence in democracy slowly, but confidence does spread. The bureaucrat with a secure, democratic home life can be expected to feel more confident bending rules with clients. Eventually, as in Norway, it becomes as unacceptable for a prime minister to advocate—let alone legislate—corporal and capital punishment alike. Gradually, micro-changes can metastasize and be reflected in macro-changes.

Is democracy worth the personal commitment if macro-progress probably comes by millennial inches? Why shouldn't I just accept that I will leave the world about as violent as I found it, and play to win while I'm around? Why work so hard and suffer so much frustration if there's no big payoff? Isn't that asking for an awful lot of altruism?

I don't think so. Looking ahead, I'd much rather die surrounded by friends than by would-be heirs to a giant inheritance. The only reward I have is that democracy gives me and others satisfaction in itself.

On one level the theory of violence and democracy I have offered here is value-neutral. It doesn't tell anyone whether to respond to violence with violence or democracy; it only tells you what to expect to get in return when you make the choice.

On another level, this theory is heavily value laden. If the theory is worth examining and testing, it is only because democracy is worth trying. We have plenty of experience in how to do violence, and the theory tells us it is fruitless to try to win peace through violence. For those who believe in living by violence, this theory tells them nothing worth knowing.

From moment to moment, it is a profoundly religious choice whether to commit to violence or to democracy. It is a matter of the heart, of deciding what the ultimate meaning of one's existence is to be. No theory, no voice of pure reason can resolve this dilemma. That is why individual decisions whether to respond violently or democratically are in principle unpredictable. None of us makes a pure commitment to the one side or the other. The best we can aspire to for a more peaceful world is that somehow the balance in each of us, between investment in our violent or in our peaceful sides, shifts gradually toward more effort at and contemplation of democracy. Profound though the choice may be, we reevaluate the choice from moment to moment, from context to context. In that profound malleability lies my hope, though by no means my prophecy, that people will find democracy worth studying and trying.

For me, it is a relief to discover a way to be a criminologist without being preoccupied with how to hurt or fight people. It's a relief to be able to talk about how people make peace. I have so little hope about punishing ourselves out of crime. I am so cynical about the motives of politicians who declare wars on crime. It is a great comfort to enter a realm of human interaction where this nonsense doesn't occur. And after returning from Norway and beginning to notice democracy, the more I look for it, the more I find it, and more (and yet smaller) opportunities I find to try to nurture it along.

I came to criminology believing that crime was a behavior, and trying to find it and define it. I now understand that crime is at root a relationship among human spirits. We may at any time and place proscribe behavior because we impute invariant motives to actors and those they act toward, but these legal conventions are at best imperfect approximations of the defects in relationships among human motives that truly concern us. With the understanding that the interaction of motives is the real issue, political ambiguities of defining crime are wiped away. Only then can the fundamental antithesis of crime be given form, only then can life free of crime be articulated and planned.

At best, this theory is only a starting point for studying and understanding the difference between crime and peace. Whether people really cool down and gain satisfaction from democratic forms of interaction remains open to further inquiry and test. Much can be discovered about democracy as against violence at the micro-level. At the macro-level, however, whether anyone's commitment here and now to democracy pays off is bound to remain indeterminate, while historically micro-level data remain elusive. One might imagine that if we began filling our journals with micro-analyses today, researchers centuries hence could examine whether micro-trends generated macro-trends. Meanwhile, odds are that no human being has enough years on this planet to die knowing whether—cycles of violence aside—the larger world is a more violent or a more peaceful place than it was when the person entered it. Considerable room for profoundly religious speculation whether violence pays will almost surely remain. Most of us will die convinced that the truth we have invested in the more heavily in our lives remains the truth, and frightened that those who survive us don't know it.

My aim for this book is modest: that a few readers find it more possible to take a few steps to think and act their way out of the morass of crime and violence which surrounds us. And I'd like to die thinking people are still learning.

NOTES

1. This chapter is a revision of an article which appeared in *Justice Quarterly* 5 (December 1988): 539–63. I am indebted to people who have given me substantial criticism, advice, and encouragement in writing several drafts of this paper. They range from criminologists and peace researchers across Scandinavia to colleagues and friends from coast to coast in North America, not least the editor and referees from *Justice Quarterly*. So many people have helped me greatly that I'm boggled at the thought of naming all of them and am loath to name some without naming the others. I hope these many helpers will accept my thanks and forgive my failure to identify them by name.

2. In English the suffix *-ible* means "able to . . ."; hence "responsible" means "able to respond." The English term describes a capacity to act. Norwegian has a comparable suffix: *-bar,* as in *brukbar,* "usable." This suffix is not found in *ansvar/ansvarlig*. In the Norwegian term, *an-* means "to" or "toward," while *svar* means "answer," "response," or "correspondence."

3. See Brock-Utne (1985) and Galtung (1969) for comprehensive surveys of what people call "violence."

4. Christie calls them "pain-inflicting" and "pain-free" communities.

5. Fortunately there are many exceptions which confound even the best predictors. Otherwise change would be impossible.

6. This is a failing of empirical tests to determine whether inequality causes violence. Inequality and violence are fraternal offspring, twins of an underlying unresponsiveness; neither one causes the other.

7. The same principle has led me to advocate that social scientists seek to discover ways to increase the unexplained variance in additive models of human behavior (Pepinsky 1982c).

8. The original version of this chapter appeared in the *Australian Journal of Law and Society* 4 (1988): 42–60.

9. The original version of this chapter appeared in *Crime and Delinquency* 35 (July 1989): 458–70.

REFERENCES

Aftenposten (1986). "Mangelfull Informasjon: 16 av 129 Visste Ikke Hva De Ble Operert For" ("Insufficient Information: 16 of 129 Did Not Know What They Were Operated on For"). February 24, p. 18.

Andenaes, Johannes (1966). "The General Preventive Effect of Punishment." *University of Pennsylvania Law Review* 114 (May): 949–83.

Aubert, Vilhelm (1959). "The Role of Chance in Social Affairs." *Inquiry* 2 (Spring 1959): 1–24.

Bateson, Gregory (1979). *Mind and Nature: A Necessary Unity.* New York: Dutton.

Batra, George (1987). *The Great Depression of 1990: Why It's Got to Happen—How to Protect Yourself.* New York: Simon and Schuster.

Bayley, David H. (1976). *Forces of Order: Police Behavior in Japan and the United States.* Berkeley: University of California Press.

Beccaria, Cesare (1968 [1764]). *On Crimes and Punishments.* Indianapolis: Bobbs-Merrill.

Brantingham, Patricia L., and Paul J. Brantingham (1975). "Residential Burglary and Urban Form." *Urban Studies* 12 (October): 273–84.

Breeden, William T. (1987). As quoted in the Anniston (Alabama) *Star,* November 24, p. 5.

Brock-Utne, Birgit (1985). *Educating for Peace: A Feminist Perspective.* Elmsford, N.Y.: Pergamon.

Brogden, Michael, with A. Brogden (1982). *The Police: Autonomy and Consent.* London: Academic Press.

Cain, Maureen. (1973). *Society and the Policeman's Role.* London: Routledge and Kegan Paul.

Capra, Fridtjof (1974). *The Tao of Physics: An Exploration of the Parallels Between Modern Physics and Eastern Mysticism.* New York: Bantam Books.

Chambliss, William J. (1988). *On the Take* (2d edn.). Bloomington: Indiana University Press.

——— (1987). The politics of organized crime. Unpublished but videotaped keynote lecture for the Fiftieth Anniversary of Criminal Justice at Indiana University celebration, Bloomington, March 27.

Christie, Nils. (1982). *Hvor Tett et Samfunn? (How Closely Knit a Society?)* Oslo: Oslo University Press.

——— (1981). *Limits to Pain.* Oxford, UK: Martin Robertson.

——— (1987). "Conflicts as Property." *British Journal of Criminology* 17 (January): 1–19.
Cizankas, Victor I. (1973). The future role of police. Taped oral presentation supplementing a paper at the Second Criminal Justice Conference, November 3. Santa Barbara: Center for the Study of Democratic Institutions.
Clark, John P. (1965). "The Social Isolation of the Police: A Comparison of British and American Situations." *Journal of Criminal Law, Criminology, and Police Science* 56: 307–19.
Clavelle, Peter. (1986). "A City Supports Worker Ownership: Burlington, Vermont." *Changing Work* 3: 12–14.
Clinard, Marshall B. (1978). *Cities with Little Crime: The Case of Switzerland.* New York: Cambridge University Press.
Craigmillar Festival Society. 1987. *Annual Report 1986/87.* Edinburgh, Scotland: Craigmillar Festival Society.
Crummy, H. (1987). "Craigmillar: self treatment—the cooperative approach." Unpublished paper delivered at the International Conference on Solutions to the Pathologies of Urban Processes, Kazimierz Dolny, Poland, October 12–16.
Curran, Peggy (1988). "Unless your name's Tremblay this 3-day blast is not for you." *Montreal Gazette,* June 25, A-7.
Denzin, Norman K. (1982). "Notes on Criminology and Criminality." Pp. 115–30 in Harold E. Pepinsky (ed.), *Rethinking Criminology.* Beverley Hills: Sage.
Durkheim, Emile (1968 [1895]). *Rules of the Sociological Method.* New York: Free Press.
——— (1968 [1893]). *Division of Labor in Society.* New York: Free Press.
Falchenberg, Knut, Haakon Letvik, and Knut Snare (1986). "Tre Fanger: Soning Uten Innhold" ("Three Prisoners: Doing Time without Content"). *Aftenposten,* February 7, p. 49.
Foucault, Michel (A. Sheridan, trans.) (1977). *Discipline and Punish: The Birth of the Prison.* New York: Pantheon.
French, Marilyn (1985). *Beyond Power: On Women, Men and Morals.* New York: Summit Books.
Fuller, R. Buckminster (1975/1979). *Synergetics: Explorations in the Geometry of Thinking* (2 vols.). New York: Macmillan.
Galtung, Johan (1969). "Violence, Peace, and Peace Research." *Journal of Peace Research* 6, 2:167–91.
Gleick, James (1987). *Chaos: Making a New Science.* New York: Viking Penguin.
Grossman, David (H. Watzman, trans.) (1988). *The Yellow Wind.* New York: Farrar, Straus, and Giroux.
Gutierrez-Johnson, A. (1984). "The Mondragon Model of Cooperative Enterprise: Conditions Concerning Its Success and Transferability." *Changing Work,* 1:35–41.
Harris, M. Kay (1985). "Toward a Feminist Vision of Justice." Un-

published paper presented at Amsterdam, June, Second International Conference on Prison Abolition.

Henk, Thomas, and Chris Logan (1982). *Mondragon: An Economic Analysis*. London/Boston: Allen & Unwin.

Henry, Stuart (1984). *Private Justice: Towards Integrated Theorizing in the Sociology of Law*. New York: Methuen.

Hough, M., and P. Mayhew (1985). *Taking Account of Crime: Findings from the 1984 British Crime Victim Survey*. London: Her Majesty's Stationery Office.

Hulsman, Louk H. C. (1986). "Critical Criminology and the Concept of Crime." *Contemporary Crises* 10 (July): 63–80.

Ivanicka, K. (1988a). *Synergetika a civilizacia*. Bratislava: Alfa.

——— (1988b). "Dynamics of the system 'res publica' on the development of self-government from 'obcina' to federation in Czechoslovakia." Unpublished paper given at the "Res Publica: East and West" Conference, Dubrovnik, October 19–14.

Jacobs, Jane (1961). *The Death and Life of Great American Cities*. New York: Random House.

Jesilow, Paul (1982). "Adam Smith and the Notion of White-Collar Crime: Some Research Themes." *Criminology* 20 (November): 319–28.

Johansen, Per Ole (1984). *Oss Selv Naermest (Norwegians First)*. Oslo: Gyldendal.

Klein, Alexander (ed.) (1969). *Natural Enemies??? Youth and the Clash of Generations*. Philadelphia: Lippincott.

Kramer, Ronald C. (1982). "From 'Habitual Offenders' to 'Career Criminals': The Historical Development of Criminal Categories." *Law and Human Behavior* 6 (nos. 3/4): 273–93.

Latimer, George (1986 [1983]). "What Makes a Self-Reliant City?" *Changing Work* 2 (Spring): 10.

Letvik, Haakon (1986) "Insatte og Ansatte paa Ila: Skjerpet Narkokontroll Skadelig" ("Inmates and Staff at Ila [Prison]: Sharpened Drug Control Harmful"). *Aftenposten*, June 5, p. 13.

McNamara, J. H. (1967). "Uncertainties in police work." In D. J. Bordua, (ed.), *The Police: Six Sociological Essays*. New York: Wiley.

Marongiu, Pietro, and Graeme Newman (1987). *Vengeance: The Fight Against Injustice*. Totowa, N.J.: Rowman and Littlefield.

Mathiesen, Thomas (1986). "The Politics of Abolition." *Contemporary Crises* 10 (July): 81–94.

Melossi, Dario, and Massimo Pavarini (G. Cousin, trans.) (1981). *The Prison and the Factory: Origins of the Penitentiary System*. Totowa, N.J.: Barnes and Noble.

Merry, Sally Engle (1981). *Urban Danger*. Philadelphia: Temple University.

Milgram, Stanley (1973). *Obedience to Authority: An Experimental View*. New York: Harper and Row.

Mills, C. Wright (1956). The Power Elite. New York: Oxford University Press.

Monahan, John (1978). "Prediction Research and the Emergency Commitment of Dangerous Mentally Ill Persons: A Reconsideration." *American Journal of Psychiatry* 135 (February): 198–201.

Montero, Jorge (1983). "Costa Rica." Pp. 233–49 in Elmer H. Johnson (ed.), *International Handbook of Contemporary Developments in Criminology: General Issues and the Americas.* Westport, Conn.: Greenwood Press.

Morris, David. (1986). "The Self-Reliant City: St. Paul." *Changing Work* 3: 8–11.

Moyers, Bill (1988). *The Secret Government.* Cabin John, Md.: Seven Locks Press.

Murton, Thomas O., and Joe Hyams (1970). *Accomplices to the Crime.* New York: Grove Press.

Newman, Graeme, and Michael Lynch (1987). "From Feuding to Terrorism: The Ideology of Vengeance." *Contemporary Crises* 11 (no. 3): 223–42.

Nyerere, Julius K. (1969). "Socialism and Rural Development." Pp. 246–71 in Knud Erik Svendsen and Merete Teisen (eds.), *Self-Reliant Tanzania.* Dar es Salaam: Tanzania Publishing House.

Parnell, Philip C. (1988). *Escalating Disputes: Social Participation and Change in the Oaxacan Highlands.* Tucson, Arizona: University of Arizona Press.

Pepinsky, Harold E. (1987a). "Explaining Police-Recorded Crime Trends in Sheffield." *Contemporary Crises* 11, 1:59–73.

——— (1987b). "Information Sharing as a Human Right." *Humanity and Society* 11 (May): 189–211.

——— (1985). "A Sociology of Justice." Pp. 93–108 in Ralph Turner (ed.), *Annual Review of Sociology.* Palo Alto: Annual Reviews.

——— (1984). "Better Living Through Police Discretion." *Law and Contemporary Problems* 47 (Fall): 249–67.

——— (1982a). "A Season of Disenchantment: Trends in Chinese Justice Reconsidered." *International Journal of the Sociology of Law* 10 (August): 277–85.

——— (1982b). "Toward a Science of Confinement, Away from the Fallacy of the Counterroll, in Criminology." Pp. 35–45 in Harold E. Pepinsky (ed.), *Rethinking Criminology.* Beverly Hills: Sage.

——— (1982c). "Humanizing Social Control." *Humanity and Society* 6 (August): 227–42.

——— (1980). *Crime Control Strategies.* New York: Oxford.

——— (1978). "Communist Anarchism as an Alternative to the Rule of Criminal Law." *Contemporary Crises* 2 (July): 315–34.

——— (1977). "Stereotyping as a Force for Increasing Crime Rates." *Law and Human Behavior* 1 (September): 299–308.

——— (1976a). *Crime and Conflict: A Study of Law and Society.* Oxford: Martin Robertson.

——— (1976b). "Police Patrolmen's Offense-Reporting Behavior." *Journal of Research in Crime and Delinquency* 13 (January): 33–47.

——— (1975a). "Police Decision-Making." Pp. 21–52 in Don M. Gottfredson (ed.), *Decision-Making in the Criminal Justice System: Review and Essays.* Washington, D.C.: U.S. Government Printing Office.

——— (1975b). "Reliance on Formal Written Law, and Freedom and Social Control, in the United States and the People's Republic of China." *British Journal of Sociology* (September): 330–42.

Pepinsky, Harold E., and Paul Jesilow (1985). *Myths That Cause Crime.* Revised and annotated edition. Cabin John, Md: Seven Locks.

Pepinsky, Pauline Nichols (1961). "The Social Dialectic of Productive Nonconformity." *Merrill-Palmer Quarterly of Behavior and Development* 7 (April): 127–37.

Pines, Maya (1979). "Superkids." *Psychology Today* 12 (January): 52–58.

Quinney, Richard (1987). The Way of Peace: Notes for a New Criminology. Paper presented at meeting of the American Society of Criminology, Montreal.

——— (1982). *Social Existence: Metaphysics, Marxism, and the Social Sciences.* Beverly Hills: Sage.

Regoli, Bob, and John D. Hewitt (1991). *Delinquency in Society.* New York: Macmillan (in press).

Reiman, Jeffrey (1984). *The Rich Get Richer and the Poor Get Prison.* 2d edn. New York: Wiley.

Rosenbaum, D. P. (1988). "Community Crime Prevention: A Review and Synthesis of the Literature." *Justice Quarterly* 5: 323–95.

Ruddick, Sara (1989). *Maternal Thinking: Toward a Politics of Peace.* Boston: Beacon Press.

——— (1980). "Maternal Thinking." *Feminist Studies* 6 (Summer): 342–67.

Schumacher, E. F. (1975). *Small Is Beautiful: Economics As If People Mattered.* New York: Harper and Row.

Schütz, Alfred (1970). *On Phenomenology and Social Relations.* Chicago: University of Chicago Press.

Selke, William L. (1989). "The opening of prisons." Unpublished paper presented at the Fourth International Conference on Penal Abolition, Kazimierz Dolny, Poland, May.

Selke, William L., and Harold E. Pepinsky (1982). "Police Reporting in Indianapolis, 1948–1978." *Law and Human Behavior* 6, 3/4:327–42.

Shared Housing Quarterly (published quarterly). Philadelphia: Shared Housing Resource Center, Inc.

Shoham, S. Giora (1985). *Rebellion, Creativity and Revelation.* New Brunswick, N.J.: Transaction Books.

Smith, Adam (1937 [1776]). *The Wealth of Nations.* New York: Modern Library.

Smykla, John Ortiz (1987). "The Human Impact of Capital Punishment: Interviews with Families of Persons on Death Row." *Journal of Criminal Justice* 15, 4: 331–47.

Sonnleitner, Michael W. (1989). "Gandhian Satyagraha and Swaraj: A Hierarchical Perspective." *Peace and Change* 14 (January): 3–24.

Stang Dahl, Tove (1987). "Women's Law: Methods, Problems, Values." *Contemporary Crises* 10, 4: 361–71.

———, ed. (1985). *Kvinnerett I & II* (*Women's Law,* Vols. 1 and 2). Oslo: Universitetsforlaget (University Press).

Sudnow, David (1965). "Normal Crimes: Sociological Features of the Penal Code in a Public Defender Office." *Social Problems* 12 (Winter): 255–76.

TANU (Tanganyika African National Union) (1967). "The Arusha Declaration." Appears at pp. 184–200 in Knud Erik Svendsen and Merete Teisen (eds.), *Self-Reliant Tanzania.* Dar es Salaam: Tanzania Publishing House (1969).

Tierney, John, Lynda Wright, and Karen Springer (1988). "The Search for Adam and Eve." *Newsweek* 111 (Jan. 11): 46–52.

Tifft, Larry L. (1985). "Reflections on Capital Punishment and the 'Campaign Against Crime' in the People's Republic of China." *Justice Quarterly* 2 (March): 127–37.

Tifft, Larry, and Dennis Sullivan (1980). *The Struggle to Be Human: Crime, Criminology and Anarchism.* Over the Water, Sanday, Orkney, UK: Cienfuegos.

Tocqueville, Alexis de (1956 [1840]) Richard D. Heffner, ed.) *Democracy in America.* New York: New American Library.

———(1945 [1840]). "What Sort of Despotism Democratic Nations Have to Fear." Book 4, chap. 3, at pp. 334–39, in *Democracy in America,* Vol. 2. New York: Vintage Books.

Van Velzen, H. U. E. T., and W. Van Wetering (1960). "Residence, Power Groups, and Intra-societal Aggression: An Enquiry into the Conditions Leading to Peacefulness within Non-stratified Societies." *International Archives of Ethnography* 49: 169–200.

Weber, Max (1946 [1918]). "Politics as a Vocation" and "Science as a Vocation," Pp. 77–128, 129–56 in Hans H. Gerth and C. Wright Mills (eds. and trans.), *From Max Weber: Essays in Sociology.* New York: Oxford University Press.

Wilkins, Leslie T. (1984). *Consumerist Criminology.* London: Heinemann.

Witcher, Nan (1986). "The Captain and the Cop." *ACJS Today* (September): 4–5.

Witty, C. (1980). *Mediation and Society: Conflict Management in Lebanon.* New York: Academic Press.

Zimbardo, Phillip G. (1978). "The Psychological Power and Pathology of Imprisonment." Pp. 202–10 in John R. Snortum and Ilana Hadar (eds.), *Criminal Justice: Allies and Adversaries.* Pacific Palisades, Cal.: Palisades.

Zinn, Howard (1980). *A People's History of the United States.* New York: Harper and Row.

Zuñiga, Ricardo B. (1975). "The Experimenting Society and Radical Social Reform: The Role of the Social Scientist in Chile's Unidad Popular Experience." *American Psychologist* 30: 90–115.

HAROLD E. PEPINSKY is Professor of Criminal Justice and of East Asian Languages and Cultures, Indiana University, Bloomington. He is author of *Crime and Conflict: A Study of Law and Society* and *Crime Control Strategies* and co-author with Paul Jesilow of *Myths That Cause Crime*. He is editor of several volumes, including the companion volume to this, co-edited with Richard Quinney, *Criminology as Peacemaking*.